PROVIDING FOR NATIONAL SECURITY

PROVIDING FOR NATIONAL SECURITY

A Comparative Analysis

Edited by Andrew M. Dorman
and Joyce P. Kaufman

Stanford Security Studies
An Imprint of Stanford University Press
Stanford, California

Stanford University Press
Stanford, California

Printed in the United States of America on acid-free, archival-quality paper

Library of Congress Cataloging-in-Publication Data

Providing for national security : a comparative analysis / edited by Andrew M. Dorman and Joyce P. Kaufman.
pages cm
Includes bibliographical references and index.
ISBN 978-0-8047-9066-6 (cloth : alk. paper)
ISBN 978-0-8047-9155-7 (pbk. : alk. paper)
1. National security. 2. World politics—1989– I. Dorman, Andrew M., 1966– editor of compilation. II. Kaufman, Joyce P., editor of compilation.
UA10.5. P76 2014
355'.0335—dc23

2013040087

Contents

Preface

THIS VOLUME FOLLOWS ON FROM OUR EARLIER PROJECT ON *The Future of Transatlantic Relations* (Stanford: Stanford University Press, 2011) and examines the burgeoning area of national security. Over the last decade or so, this area of public policy has received far greater prominence as policy-makers, think-tanks, and the academic community have sought to redefine security and consider new ways of providing for national security. The thirteen case studies contained within this volume individually and collectively provide a fascinating insight into the national security process and help show how factors such as culture, geography, and history play major parts in the policy process. Involving country experts has proven to be an extremely fruitful advantage in helping us to understand where individual states were coming from and often where they also aspired to head toward. For us, the comparative analysis provided insights that we were not expecting, particularly similarities between states that we had not seen as obvious comparators.

The editors would like to thank all the contributors to this volume for their willingness to draft and revisit each of their chapters, meet deadlines, and provide input in producing what we believe is an edited collection that provides some real insights. We would also like to thank Geoffrey Burn and his team at Stanford University Press. Their professionalism has meant that we have found, once again, their support and advice invaluable and the process of bringing an edited collection together relatively trouble-free.

Andrew M. Dorman, Oxford, England
Joyce P. Kaufman, Whittier, CA

Notes on Contributors

Robert H. Donaldson is Trustees Professor of Political Science at the University of Tulsa, where he was president from 1990 to 1996. Previously, he was president of Fairleigh Dickinson University; provost of Lehman College of the City University of New York; and a professor and associate dean at Vanderbilt University. He is a member of the Council on Foreign Relations and was a Council International Affairs Fellow from1973–1974, serving as a consultant with the U.S. Department of State; he was also a visiting research professor at the Strategic Studies Institute, U.S. Army War College from 1978–1979. He is author or coauthor of six books and more than two dozen articles and book chapters, primarily on the politics and foreign policy of the USSR and Russia; his most recent book is *The Foreign Policy of Russia: Changing Systems, Enduring Interests*, 4th ed., 2009.

Andrew M. Dorman is a Professor of International Security at King's College London and an Associate Fellow of the International Security Programme at the Royal Institute of International Affairs, Chatham House. His research focuses on decision-making and the utility of force, utilizing the case studies of British defense and security policy and European Security. His recent books include coediting (with Joyce Kaufman) *The Future of Transatlantic Relations: Perceptions, Policy and Practice*, (Palo Alto, Calif.: Stanford University Press, 2011); *Blair's Successful War: British Military Intervention in Sierra Leone* (Farnham, UK: Ashgate, 2009). He originally trained as a chartered accountant with the professional company KPMG, qualifying in 1990 before returning to academia. He has previously taught at the University of Birmingham, where he completed his masters and doctoral degrees and the Royal Naval College Greenwich.

Jon Hill is a Senior Lecturer in the Defence Studies Department at King's College London, UK. He has published widely on issues of African security. His main publications include *Nigeria Since Independence: Forever Fragile?*; *Identity in Algerian Politics: The Legacy of Colonial Rule*; *Remembering the War of Liberation: Legitimacy and Conflict in Contemporary Algeria*; *Sufism in Northern Nigeria: A Force for Counter-Radicalisation?*; *Islamism and Democracy in the Modern Maghreb*; *Corruption in the Courts: The Achilles Heel of Nigeria's Regulatory Framework?*; and *Thoughts of Home: Civil-Military Relations and the Conduct of Nigeria's Peacekeeping Forces.*

Chris Hughes is Professor of International Politics and Japanese Studies, chair of the Department of Politics and International Studies, and chair of the Faculty of Social Sciences at the University of Warwick, UK. He was formerly research associate at the University of Hiroshima; *Asahi Shimbun* visiting professor of mass media and politics, University of Tokyo; and Edwin O. Reischauer visiting professor of Japanese studies, Department of Government, Harvard University. He holds adjunct positions at Hiroshima, Waseda, and Harvard universities. His most recent publications include *Japan's Remilitarisation* (Routledge, 2009), and *Japan's Reemergence as a "Normal" Military Power* (Oxford University Press, 2004). He is currently president of the British Association of Japanese Studies and joint editor of *The Pacific Review.*

Joyce P. Kaufman is a Professor of Political Science and Director of the Center for Engagement with Communities at Whittier College. She is the author of *Introduction to International Relations* (2013), *A Concise History of U.S. Foreign Policy*, 2nd ed. (2010), and *NATO and the Former Yugoslavia: Crisis, Conflict and the Atlantic Alliance* (2002) and coeditor of *The Future of Transatlantic Relations: Perceptions, Policy and Practice* (with Andrew M. Dorman) (2011). She is also the author of numerous articles and papers on U.S. foreign and security policy. With Kristen Williams, she is coauthor of *Challenging Gender Norms: Women and Political Activism in Times of Crisis* (2013), *Women and War: Gender Identity and Activism in Times of Conflict* (2010), and *Women, the State, and War: A Comparative Perspective on Citizenship and Nationalism* (2007).

Maryanne Kelton is a Senior Lecturer in international relations with the School of International Studies and deputy director of the Centre for United States and Asia Policy Studies at Flinders University. Her research interests concern the Australia–U.S. relationship, Australian foreign policy, defense procurement, and economic statecraft. She is author of *"More than an Ally"? Contemporary Australia-US Relations* (Ashgate, 2008) and *New Depths: The Collins Class Submarine Project* (ANU, 2005). She was also coauthor of the post-conflict transition simulation for the International Peace and Security Institute's The Hague Symposium in 2012. Among other scholarly papers and book chapters, she has also published in the *International Relations of the Asia-Pacific, Australian Journal of International Affairs,* and *Australian Journal of Politics and History.*

Gale A. Mattox is a Professor of Political Science, U.S. Naval Academy; recipient of USNA Superintendent's Civilian Faculty Service Award; adjunct professor, Graduate Security Studies, Georgetown University; chair, International Security & Arms Control Section, APSA; visiting senior scholar for Foreign/Domestic Policies, AICGS, Johns Hopkins University, and 2009 Distinguished Fulbright-Dow Research Chair, Roosevelt Center, The Netherlands. She served on the U.S. Department of State Policy Planning Staff and in Office Strategic and Theater Nuclear Policy; was International Affairs Analyst, Congressional Research Service; president, Women in International Security (WIIS), Georgetown University, and vice president, International Studies Association. She is on the boards of George Marshall Center, Germany; American Friends of the Alexander von Humboldt Foundation; and Forum for Security Studies, Swedish National Defense University. Recent publications include "Tactical Nuclear Weapons in Europe," *Arms Control History, Theory, and Policy* (2012), Williams/Viotti (eds.); "Resetting the U.S.–Russian Relationship: Is 'Cooperative Engagement' Possible?," *European Security* (2011); "Germany: From Civilian Power to International Actor," *The Future of Transatlantic Relations* (2011), Dorman/Kaufman (eds.), Stanford University Press.

Patrick M. Morgan is a Professor of Political Science and holds the Tierney Chair in Global Peace and Conflict Studies at the University of California, Irvine. He is a founding and board member of the Council on U.S.–Korean Security Studies. His specialties are national and international security, U.S. security relations with Northeast Asia, and deterrence theory. His latest book (coedited) is *Complex Deterrence* (University of Chicago Press, 2009), and his current research project is the evolution of the American alliance system since the Cold War.

Harsh V. Pant is a Reader in International Relations in the Defence Studies Department at King's College London. He is also an Associate at the Centre for Science and Security Studies and the India Institute at King's College London. He is also an adjunct fellow with the Wadhwani Chair in U.S.–India Policy Studies at the Center for Strategic and International Studies, Washington, D.C. He has been a visiting professor at the Indian Institute of Management, Bangalore; a visiting fellow at the Center for the Advanced Study of India, University of Pennsylvania; and a visiting scholar at the Center for International Peace and Security Studies, McGill University. He is presently an Emerging Leaders Fellow at the Australia-India Institute, University of Melbourne. His current research is focused on Asian security issues. His most recent books include *The China Syndrome* (HarperCollins, 2010), *The US-India Nuclear Pact: Policy, Process and Great Power Politics* (Oxford University Press, 2011), and *The Rise of China: Implications for India* (Cambridge University Press, 2012).

Bill Park is a Senior Lecturer in the Department of Defence Studies, King's College London. He has written journal articles, book chapters, and monographs on a range of Turkish foreign policy issues, including its EU accession prospects, Turkey and ESDP, policies toward Northern Iraq, Turkey–U.S. relations, the Fethullah Gulen movement, and the Ergenekon affair. In 2011, Routledge published his book on Turkish foreign policy and globalization, and he is currently working on a project on Turkey's relationship with the Kurdish regional government in northern Iraq. He has appeared as a Turkey expert on various media in the UK, Turkey, and Australia and is occasionally used as a consultant on Turkish issues by various UK government agencies.

David Rudd is a senior defense scientist with the Centre for Operational Research and Analysis (CORA) at the Department of National Defence in Ottawa. He is currently attached to the strategic analysis team in the office of the Associate Deputy Minister (Policy). His previous postings include the Canadian Forces' Operational Support Command, the Chief of Force Development's Directorate of Future Security Analysis, and the Royal Canadian Navy's Directorate of Maritime Strategy. His specialties include NATO and Canadian military transformation, NATO-EU relations, Arctic security, and the transformation of allied navies. Mr. Rudd contributed a chapter on

Canada to the Stanford Security Series volume, *The Future of Transatlantic Relations: Perceptions, Policy, and Practice* (2011).

Adrian Treacher is a Lecturer in the Department of Politics and Contemporary European Studies. Following six and a half years of postgraduate, doctoral, and postdoctoral study at the University of Birmingham (UK), Adrian Treacher has spent the last ten and a half years at the University of Sussex's prestigious Sussex European Institute. In addition to researching European security and EU external relations, he has also focused on French foreign policy with articles for *International Peacekeeping* and *European Security*, among others. He is also the author of *French Interventionism: Europe's Last Global Player?* (Ashgate, 2003).

Kathleen (Kate) Walsh is an Associate Professor of National Security Affairs at the U.S. Naval War College and affiliate of the China Maritime Studies Institute (CMSI). Her research focuses on China and the Asia-Pacific region, particularly security and technology issues. Her numerous publications include "Chinese Peacekeeping in the Asia Pacific: A Case Study on East Timor," coauthored with Lyle Goldstein, in CMSI Red Book 9 (2012); "Enhanced Information Sharing in the Asia Pacific: Establishing a Regional Cooperative Maritime Operations Center," in *Strategic Manoeuvres: Security in the Asia-Pacific*, James Veitch, ed. (2009); and *Foreign High-Tech R&D in China: Risks, Rewards, and Implications for U.S.-China Relations* (2003). Walsh is a member of the National Committee on U.S.-China Relations and the U.S. Council on Security Cooperation in the Asia Pacific (CSCAP). She has an M.A. from Columbia University and a B.A. from George Washington University.

Abbreviations

9/11	11 September 2001 terrorist attacks on the United States
ADF	Australian Defence Force
AKP	Adalet ve Kalkinma Partisi—Justice and Development Party
AMISOM	African Union Mission in Somalia
ANPP	All Nigeria People's Party
ANZUS	Australia, New Zealand, United States
AoC	Alliance of Civilizations
APEC	Asia-Pacific Economic Cooperation
AQLIM	Al Qaeda in the Land of the Islamic Maghreb
ARF	ASEAN Regional Forum
ASAT	Anti-satellite
ASEAN	Association of South East Asian Nations
ASIO	Australian Secret Intelligence Organisation
AUD	Australian Dollars
AU	African Union
AusAID	Australian Aid
AUSMIN	Australia U.S. Defence and Foreign Ministerial meetings
AUSFTA	Australia–U.S. Free Trade Agreement
BJP	Bharatiya Janata Party
BLACKSEAFOR	Black Sea Naval Cooperation Task Group
BMD	Ballistic Missile Defence
BSEC	Black Sea Economic Cooperation

CCP	Chinese Communist Party
CDS	Chief of Defence Staff
CIS	Commonwealth of Independent States
CSTO	Collective Security Treaty Organization
DG	Director-General
DRC	Democratic Republic of the Congo
ECOWAS	Economic Community of West African States
EU	European Union
EurAsEC	Eurasian Economic Community
FSB	Federal Security Service
GFC	Global financial crisis
GDP	Gross domestic product
GoM	Group of Ministers
IAEA	International Atomic Energy Agency
IBB	Ibrahim Badamasi Babangida
ICBM	Intercontinental Ballistic Missile
IFOR	Implementation force
IMF	International Monetary Fund
IMN	Islamic Movement in Nigeria
INTERFET	International Force for East Timor
ISAF	International Security Assistance Force
JTF	Joint task force
KRG	Kurdish regional government
LTIPP	Long-term integrated perspective plan
MEND	Movement for the Emancipation of the Niger Delta
NATO	North Atlantic Treaty Organization
NDA	National Democratic Alliance
NDPVF	Niger Delta Peoples Volunteer Force
NDVF	Niger Delta Vigilante Force
NICC	National Intelligence Coordination Committee
NSA	National Security Advisor
NSC	National Security Council
NSG	Nuclear Suppliers Club
NSS	National Security Statement/Strategy
OAU	Organization of African Unity
OIC	Organization of Islamic Conference
PfP	Partnership for Peace

PKK	Partiya Karkerên Kurdistan—Kurdistan Workers' Party
PLA	People's Liberation Army
PRC	People's Republic of China
PRT	Provincial reconstruction team
PSO	Peace Support Operations
RAMSI	Regional Assistance Mission Soloman Islands
RF	Russian Federation
S&T	Science and technology
SCO	Shanghai Cooperation Organization
SFOR	Stabilization force
SLOCs	Sea Lanes of Communications
SSS	State security service
TGNA	Turkish Grand National Assembly
UK	United Kingdom
UKUSA	United Kingdom United States
UN	United Nations
UNCD	United Nations Conference on Disarmament
UNCLOS EEZ	United Nations' Convention on the Law of the Sea Exclusive Economic Zone
UNIFIL	UN Interim Force in Lebanon
UNPROFOR	UN Protection Force
UNSC	United Nations Security Council
US	United States
USSR	Union of Soviet Socialist Republics
WMD	Weapons of mass destruction
WTO	World Trade Organization
WW2	World War 2

PROVIDING FOR NATIONAL SECURITY

PART I

INTRODUCTION

1 The Challenge of National Security

Andrew M. Dorman and Joyce P. Kaufman

IN THE SECOND DECADE OF THE 21ST CENTURY, THE TREND CONTINues for an increasing number of states to publish some form of national security strategy.[1] Yet what comprises "national security" remains no less contentious today than when Arnold Wolfers identified the ambiguities within it in the 1950s or when Barry Buzan, borrowing W.B. Gallie's phrase, described national security as an "essentially contested concept" or when Peter Katzenstein brought together a number of scholars together to examine the role of culture on national security.[2]

Those responsible for the provision of or engaged in teaching on national security are confronted by several basic questions:

- Who provides national security? Is it the sole preserve of the state or has some of that responsibility moved toward international organizations such as the United Nations, NATO, and the International Fund, or is responsibility increasingly passing to the private sector?
- Who is national security provided for? Is the ultimate role of the provider the protection of the provider and its instruments of provision or is it about the provision of security for those who the provider is responsible for? In the case of the state, is it the state itself, the organs of the state, or the people?
- How should it be provided? What are the most appropriate tools to employ? How are decisions to be made about relative prioritization?

While the questions might initially appear quite straightforward, there is a background of potential changes and challenges to the current international order that includes a considerable debate over the utility of force. In the aftermath of the wars in Afghanistan and Iraq, the likelihood of such large-scale interventions, at least in the near term, seems remote. However, the prospect for the limited use of force, discrete military operations, or whatever nomenclature is chosen, still remains.[3] The French deployment of forces to Mali in 2013 and the NATO-led operation in Libya in 2011 suggest that the impulse toward liberal interventionism may not be dead but that the will to affect the means to intervene is running behind both the stated aim and the political rhetoric.[4] In both these examples, the United States, under President Obama, has taken a back seat, and rather than leading has allowed other countries to take the lead in addressing global security issues. This does not mean that the United States has not been willing to engage, for they did in both cases, but in a support role to the European allies. The question that rises out of this is, why? How does this link to traditional ideas of national security?

The rise of Brazil, India, and China and the reemergence of Russia pose a challenge to the established post–Cold War order, not just of the dominance of the West and in particular the United States, but Western rules, norms, and values on intervention, sovereignty, and the rules of war.[5] Will a post-American world, post-Western world resemble the past in some way, such as the 19th century Concert of Europe, or in what ways will these new powers shift or set the terms of intervention and warfare in a totally different direction? Are we witnessing the inevitable decline of the United States and the emergence of Asia rather than Europe?[6] Or, concomitantly, are we witnessing the rise of other hegemonic powers that will dominate their own regions, if not the international system as a whole?[7] China's response to the reduction in the United States' credit rating, which involved a call for a new more stable currency to replace the U.S. dollar, seems to indicate that the global balance of power is changing to the detriment of the West.[8] Added to this has been the so-called Arab Spring or Arab awakening, which has seen internal challenges to a number of North African and Middle Eastern states with some regimes falling and others engaged in the systematic suppression of internal opposition.

Fear of the use of weapons of mass destruction (WMD), either by terrorists or states, reentered the academic and political discourse. Despite its failure to find WMDs, the United States led a coalition of states into Iraq in 2003, at least in part motivated by the need to remove the perceived

WMD threat through a preventive war.[9] Subsequently, Israel attacked and destroyed a potential nuclear plant in Syria.[10] Debate surrounds the possibility of an Israeli/American strike on Iran's nuclear facilities.[11] Moreover, throughout the Middle East, many states are considering acquiring their own nuclear arsenals in response to the possibility of a nuclear-armed Iran, and as a result the discussion of the modalities of limited nuclear war has returned to policy debate. These events have played out against the backdrop of the revolutions sweeping the Middle East, with great uncertainty as to their outcomes, which makes security planning all the more challenging. Following a third North Korean nuclear test, the merits of tactical nuclear weapons are being discussed once again, while in Europe a new debate has developed over the utility and future of NATO's tactical nuclear weapons. These come at a time when President Obama has been pushing for nuclear arms control, if not total disarmament. As a result, there are questions being asked about the traditional cost-benefit assumptions and whether WMDs might again feature as weapons of war.

There are potentially significant changes in how wars might be fought.[12] Are we seeing a so-called revolution in military affairs? If so, does this mean that the character of war is changing? What will that mean for national security? What impact will robotic use have on warfare? In the 1990s, precision-guided munitions usage made the use of force less costly domestically and facilitated an expansion in their use. Since the events of 9/11, the development and use of unmanned aerial vehicles (UAVs, more commonly referred to as drones) has expanded. What impact will such technological developments have on our conception of war and the use of force? For the Obama administration, the use of UAVs rather than manned aircraft over Libya meant that the administration did not need to go to Congress for approval under the War Powers Act.

There is also the question about the actual meaning of "conflict." The STUXNET cyber attack on Iran's nuclear facilities killed nobody but significantly affected the Iranian nuclear program.[13] It raised the prospect of similar style cyber attacks having a major impact on individual societies without necessarily killing anyone. Without casualties, does such an attack constitute an act of war? If so, are we moving towards potentially bloodless conflicts? Within NATO, there is a debate about whether such an attack would invoke Article V. Clearly, attacks of this nature raise broader questions about what types of response are appropriate, and do the responses need to be of a similar kind? Or can a state respond with military force to such attacks, even when no casualties have been directly inflicted?

The 2008 financial crisis continues to have an impact on different parts of the world. For much of Europe and to a degree in North America, austerity appears to be a phenomenon that is unlikely to disappear quickly. Conversely, other parts of the world have been less affected. Will the ongoing effects of the economic meltdown exacerbate the already changing power dynamics within the world? In some respects, 2008 represents an "East of Suez" moment—a point where a government or collection of governments accepts a change in their relative standing within the world.

Finally, the previous state monopoly of providing national security is increasingly being challenged. The credit crisis, particularly in Europe and the United States, has reinforced the position of nonstate actors such as the International Monetary Fund and the European Union in the financial markets. Moreover, the recent wave of riots in the United Kingdom has raised basic questions about the social contract between the citizen and the state.[14] This was reinforced by the tragic shooting in Norway 2011, in which a lone gunman wrought devastation amongst a future generation of political leaders. Are we likely to see the privatization of security as the state fails to provide for the needs of its citizens?

The goal of this book, therefore, is to undertake a comparative analysis of how states are approaching the formulation and implementation of national security. It adopts the premise that although states may no longer monopolize the articulation or provision of national security, they are, in general, still the main protagonists in the formulation and implementation of these policies, and it is through them that the majority of key international organizations, such as the United Nations, NATO, and the EU, work. For example, within the European Union a considerable number of sovereign powers have been handed to the European Union by its members, and those same members have retained responsibility for the provision of national defense and security. This does not mean that we exclude such organizations; rather, we will consider how they are utilized by and engage with states as part of national security. We also recognize that states are not single monolithic entities. One of the goals of this book is to compare how the state security apparatus works, and what factors influence the development and implementation of policy. The use of national security as the focus of analysis allows us to consider how states see their relative position within the international system. Thus we can examine what the drivers of these views are and what influence elements such as history, geography, and political culture play. It also allows us to consider how states are defining their national interest, to consider the extent to which this is changing, and what factors are driving these changes. We can also analyze how states are organizing themselves to provide national security, how is this being articulated in the public domain, and how much this is changing and why?

To keep this program within manageable bounds, we have identified that, apart from the United States (which is analyzed separately), there are effectively five typologies for states in terms of the national security agenda. We recognize that dividing states into these groupings is a subjective process and that there are other criteria that might be adopted for selection of groups, such as regional representation. A regional approach has strong merits in terms of global coverage, but it tends to imply that geography is the principle determinant of national security. From our review of the existing literature and discussions with potential contributors we were convinced that factors such as history, culture, and relative standing within the international system play a greater part, and we have therefore chosen to select case studies from the following groups. To try and capture the role of regional dynamics and geography, we have asked each of the contributors to consider these as some of the factors that influence the formulation and development national security. Four groups will be examined as outlined. The fifth group comprises those states that have yet to reach a position in which they can articulate national security in some form, and we have therefore chosen not to select any case studies from this group. This does not mean that they are unimportant but merely that given the constraints of time and space we believe their exclusion allows us to undertake a more balanced comparative study of the other groups. In each group, three case studies have been chosen, a number that we think reflects a balance between breadth and depth. This has given us thirteen case studies in total, not a high percentage of the world's states but one that fits within the logistical limitations of a book. The four groups are as follows.

The first group, which we have called the old world, analyzes Europe, a continent that some scholars have argued is becoming increasingly a backwater as power shifts toward the Pacific and Asia. This emphasis on relative decline within the international system essentially follows the Paul Kennedy line of the rise and fall of empires and assumes that there is little that the European countries can do, especially given the impact of austerity and the preservation of the welfare state that dominates their thinking. One would therefore assume that their individual and collective response to what is clearly a major threat to their individual interests should form a major part of their thinking on national security and that the measures taken to respond to this would also presumably be similar. Three case studies have been selected of the continent's major powers: France, Germany, and the United Kingdom. It could be argued that their selection is unrepresentative of Europe as a whole, and that would be correct, but in many respects they are the leaders of Europe and therefore they provide a good basis for comparison. All three states have significant histories that influence their current policies, all three have advanced economies that have suffered differently from the 2008 financial crisis, all are involved in many of the important international institutions (e.g., NATO, the European Union [EU], G-7) and include two permanent members of the United Nations Security Council (UNSC)—France and the United Kingdom. All three deployed military forces to Afghanistan, and two (France and the United Kingdom) are perceived have global responsibilities (e.g., Mali and Sierra Leone).

The second group, the new 20th century world, examines three case studies of nations that became important states in the international system during the 20th century—Australia, Canada, and Japan—but whose relative standing is no longer so assured. In a sense this group is where the previous group was several decades ago and therefore has a vision of what might happen to them. Each can be seen as a state that has maximized its relative position within the international system and is now faced with the prospect of this changing despite the emphasis on the Asia-Pacific region in which all find themselves. Our presumption is that each will therefore be looking to take measures to preserve their relative position and learn from the mistakes of "the old world."

The third group is called the (re-)emerging 21st century world. This comprises states that have been identified as being significant major players in the next few decades. They are, in many respects, the heirs of the preceding group and are on the upward path of relative growth and power once again. The three case studies chosen—China, India, and Russia—are regularly referred to as rising or reemerging powers within the academic and policy analysis literatures and have their own shorthand with Brazil—the BRICs. All three have the potential to alter the balance of power within their respective regions, if not globally, and all three have issues with states on their borders. Two are permanent members of the UNSC and the third—India—is one of the states most mentioned for inclusion in any reform to the UNSC. All three also have potentially significant demographic issues and a requirement to placate the expectations of their population.

The final group is the potentially (re-)emerging states. This group is effectively those states that are at an earlier stage relative to the previous group. The three case studies comprise states all of whom have been considered to be on the rise—Nigeria, South Korea, and Turkey. Whether they can maintain this ascendancy remains the subject of much speculation. Two have significant internal problems (Nigeria and Turkey) and the other has the question of its other half (North Korea) in the background. All face challenges of identity and debates about where they should focus and what they want to be.

In each of the thirteen case studies we have asked the author to consider the questions of who provides national security, who they provide it for, and how do they provide it. Given the wider factors of history, political culture, and relative standing, we have not been too prescriptive with the individual authors about their individual structures; rather, we have asked them to use their expertise to consider these questions in the form that is most appropriate for their case studies, which in itself tells us something about the case studies.

In the final section we undertake a comparative analysis of all thirteen case studies. This will be undertaken in two parts. First, a comparision of the states within each of the four typology groups is examined. Second, we will explore whether there are similarities and differences factors that span all four groups. The goal of this analysis is then to reflect on what this means for policy makers and academics in terms of our understanding of national security, the international system, and the future utility of force.

Notes

1. For a current list of relevant national security strategies and defense white papers see http://merln.ndu.edu/whitepapers.html accessed 20 February 2013.

2. Arnold Wolfers, "'National Security'" as an Ambiguous Symbol," *Political Science Quarterly* 67: 4, December 1952, p. 481; Barry Buzan, *People, States and Fear*, p. 7; W. B. Gallie, "Essentially Contested Concepts," *Proceedings of the Aristotelian Society* (Vol. 56, 1955–1956), p. 169; Peter Katzenstein (ed.), *In the Culture of National Security: Norms and Identity in World Politics* (New York: Columbia University Press, 1996).

3. Micah Zenko, *Between Threats and War: U.S. Discrete Military Operations in the Post Cold War World*, Council on Foreign Relations, 2010.

4. Andrew M Dorman, "Lessons from Libya," *Parliamentary Brief*, vol.13, no.6, April 2011, http://www.parliamentarybrief.com/2011/04/lessons-from-libya, pp.11–2, accessed 11 August 2011.

5. Sujit Dutta, "Managing and Engaging Rising China: India's Evolving Posture," *Washington Quarterly*, vol. 34, no. 2, 2011, pp.127–144.

6. Paul Kennedy, *The Rise and Fall of the Great Powers: Economic Change and Military Conflict from 1500 to 2000* (London: Fontana Press, 1988); Simon Serfaty, "Moving into a Post-Western World," *Washington Quarterly*, vol. 34, no. 2, 2011, pp. 7–23.

7. See, for example, John J. Mearsheimer, "The Gathering Storm: China's Challenge to U.S. Power in Asia," *The Chinese Journal of International Politics*, vol. 3 (2010), p. 388. Downloaded from http://mearsheimer.uchicago.edu/pdfs/A0056.pdf.

8. Leo Lewis, "A Glimpse of China in the Raw as Beijing Attacks 'Debt Addiction,'" *The Times*, 9 August 2011, p. 13.

9. Tony Blair, *A Journey* (London: Hutchinson, 2010), pp. 373–378.

10. Barbara Slavin, "Should Israel Become a 'Normal' Nation?," *Washington Quarterly*, vol. 33, no. 4, 2010, pp. 23–37.

11. Glora Eiland, "Israel's Military Option," *Washington Quarterly*, vol. 33, no. 1, 2010, pp. 115–130.

12. See, for example, Colonel John B. Alexander, *Future War: Non-Lethal Weapons in Twenty-First Century Warfare* (New York: Thomas Dunne Books, 1999); Tim Benbow, *The Magic Bullet: Understanding the Revolution in Military Affairs* (London: Brassey's, 2004); Bruce Berkowitz, *The New Face of War: How War Will Be Fought in the 21st Century* (New York: The Free Press, 2003); James R. Blaker, *Transforming Military Force* (Westport, Conn.: Praeger Security International, 2007); Risa A. Brooks and Elizabeth A. Stanley (eds.), *Creating Military Power,* (Stanford, Calif.: Stanford University Press, 2007); Robert J. Bunker (ed.), *Non-State Threats and Future Wars* (London: Frank Cass, 2003); Christopher Coker, *War in an Age of Risk* (Cambridge: Polity, 2009); Thomas J. Czerwinski, "The Third Wave: What the Tofflers Never Told You," *Strategic Forum*, no. 72 (Washington, D.C.: National Defense University, 1996); George and Meredith Friedman, *The Future of War: Power, Technology and American World Dominance in the Twenty-First Century* (New York: St. Martin's Griffin, 1996); Colin S. Gray, *Another Bloody Century* (London: Weidenfeld & Nicholson, 2005); Thomas X. Hammes, *The Sling and the Stone* (St. Paul, Minn.: Zenith Press, 2006); Michael Horowitz, *The Diffusion of Military Power: Causes and Consequences for International Politics* (Princeton, N.J.: Princeton University Press, 2010); David Kilcullen, *The Accidental Guerilla* (London: C Hurst & Co. (Publishers) Ltd., 2009); MacGregor Knox and Williamson Murray (eds.), *The Dynamic of Military Revolution 1300–2050* (Cambridge: Cambridge University Press, 2001); John Leech, *Asymmetries of Conflict: War without Death* (London: Frank Cass, 2002); Douglas A. Macgregor, *Transformation under Fire,* (Westport, Conn.: 2003); Colin McInnes, *Spectator War: The West and Contemporary Conflict* (London: Lynne Rienner, 2002); Herfried Munkler, *The New Wars* (Cambridge: Polity, 2005); John A. Nagl, *Learning to Eat Soup with a Knife: Counterinsurgency Lessons from Malaya and Vietnam* (Chicago: The University of Chicago Press, 2002); Gwyn Prins, *The Heart of War: On Power, Conflict and Obligation in the 21st century* (London: Routledge, 2002); Mikkel Vedby Rasmussen, *The Risk Society at War: Terror, Technology and Strategy in the Twenty-First Century*

(Cambridge: Cambridge University Press, 2006); Stephen Peter Rosen, *War and Human Nature* (Princeton, N.J.: Princeton University Press, 2005); General Sir Rupert Smith, *The Utility of Force: The Art of War in the Modern World* (London: Allen Lane, 2005); Alan Stephens and Nicola Baker, *Making Sense of War: Strategy for the 21st Century* (Cambridge: Cambridge University Press, 2006); Alvin and Heidi Toffler, *War and Anti-War: Making Sense of Today's Global Chaos*, (New York: Mass Market Paperback, 1995); Martin van Creveld, *The Changing Face of War* (New York: Ballantyne Books, 2007).

13. Norman Friedman, "Virus Season," *U.S. Naval Proceedings*, November 2010, pp. 88–89.

14. Tim Rayment, "England: Land of Fear and Looting," *Sunday Times*, 14 August 2011, pp. 12–15, p. 15.

2 The United States' Security Challenges of the 21st Century

Joyce P. Kaufman

Introduction

The 2012 presidential campaign was notable not for the points that it raised about the future of U.S. security policy but for what it did not. As is typical with most campaigns during a time of economic downturn, the primary emphasis was on the economy. That said, the third and final presidential debate of the campaign did focus on foreign and security policy and served as an opportunity to raise questions about the administration's judgment regarding a number of national security issues, including the handling of Iran and its nuclear weapons program; the attack in Benghazi, Libya, that resulted in the death of U.S. Ambassador Christopher Stevens and three others; and the United States' relationship with China.

The United States is facing a number of challenges, both domestic and international, that *could* alter is position and standing in the world. "*Could*" is the operative word—nothing is preordained. Despite the barrage of articles about the rise of China's power, disruptions due to the revolutions sweeping the Middle East, the economic disturbances caused by political differences over raising the debt ceiling in 2011 and then the "fiscal cliff" in 2012, and a host of other issues that the United States is confronting, there is little reason to assume that the results will be an overall decline in the status of the United States as a global power. After all, power is a relative concept, not zero-sum. The rise of one country's power need not come at the expense of another's.

This chapter will explore some of the challenges that the United States is now facing that demonstrate why the country cannot continue to pursue foreign and security policies that simply react to changing world situations rather than creating policies that move its national interest forward. Despite its unique position internationally, the United States will have to actively and deliberately redefine its foreign and national security policy to be more attuned to current realities.

Adjusting to Change

Implicit in the national security questions raised during the presidential campaign were those about the vision of America's future, and especially questions of U.S. "exceptionalism" and who could best ensure that the United States will continue to maintain its dominant role internationally. This, of course, assumes that the United States' role has declined, a debatable conclusion. One question not asked during the campaign was whether the United States *should* continue that role. Here, too, we see very different answers to that question, depending on perspective and one's assumptions about the United States.

Neo-conservatives, like Robert Kagan, point out the dominant role played by the United States in spreading democracy and in "the prolonged great-power peace" that has followed, attributing these to the "power and influence exercised by the United States . . ." He asserts that only the United States could have had this power and influence because "no other nation shares, or has ever shared, their peculiar combinations of qualities."[1] Herein lies the roots of the belief in American exceptionalism, the idea that no country except the United States could possibly have the "combination of qualities"—geographical circumstances, strong economic system, democratic form of government, military might, and of course political will—that has enabled it to play the dominant international role. The assumption is that a world that continues to be led, at least nominally, by the United States will also be a world that is more stable (peaceful). But that approach belies other realities, specifically that the world of the 21st century is not unipolar and dominated by a single hegemon but is a world with multiple power centers that will require some significant rethinking of U.S. national security policy in order to meet new challenges.

Here others provide additional insight. For example, Indyk et al. suggest that it might be more appropriate to assess Obama's policy not as a function of "success" or "failure," but more against the ideal that then-candidate, and later President Obama laid out in speeches. That ideal stands in contrast to what his administration has been able to achieve or is likely to during another term in office.[2] *The New York Times* correspondent David Sanger offers a slightly different approach when he documents and reflects on the challenges Obama faced when he inherited the policies laid out by his predecessor in Afghanistan and also Iraq. Sanger reminds us that the paramount goal then facing the new president had to be ensuring the security of the United States.[3] Thus began the odyssey that became President Obama's education in security policy leading to his somewhat controversial decisions to expand drone strikes as part of his antiterrorism offensive in Afghanistan and also Pakistan, and the use of the Stuxnet cyberworm against Iran, all pursued in the name of ensuring American security and protecting American interests.[4] That process also resulted in President Obama's decision to authorize the strike on the Bin Laden compound in Abbotabad, Pakistan, killing bin Laden but also violating the sovereignty of one of America's allies.

Despite these examples, which seem to suggest a different path for U.S. security policy, the reality is that there has been only a slight shift in policy direction since the end of World War 2, the start of the Cold War, and the emergence of the United States as a true superpower. Just as President Truman outlined a policy direction for the United States that inextricably linked military might with foreign policy, more recently President Clinton saw a globalized world in which acknowledging economic interdependence was key to U.S. success globally. Since the Clinton administration, the geographic focus might have shifted a bit from Europe to Asia and the Middle East, but the overarching direction and approaches to making U.S. national security policy remain very much the same, premised on U.S. superiority not just militarily, but economically and politically.

That said, 9/11 was a wake-up call for the United States. In the euphoria that accompanied the end of the Cold War, the United States had existed in a bubble that came with its status as "the last superpower." It is virtually impossible to understand the ways in which the United States perceives its role in the world today as well as the perceptions that other countries now have of the United States without looking at the recent past and the impact of extraordinary events, such as 9/11. Although the Bush administration engaged in a number of policy directions that were a direct response to 9/11, the Obama administration has had the advantage of time and hindsight in formulating its policies.

Going into Obama's second term with a new national security team, shifts in the direction of U.S. security policy are likely to continue. We also see a president who has a better understanding of the security climate within which he is operating, as well as the need to set priorities, understanding that it is impossible to do everything. In some ways, the challenges are greater than they were; economic realities have imposed real limits on the military, although for some critics these should have been imposed years earlier. Obama's shift in emphasis to Asia reflects geopolitical realities that are becoming more apparent and which the United States can no longer ignore.[5] The administration's reluctance to intervene in the long and bloody civil war in Syria while giving the Europeans the lead in Libya and Mali also indicates another shift in priorities away from using U.S. military force for nation-building. And the president has identified a new threat and potential form of warfare posed by cyber-terrorism.

Reflecting on the first term of the Obama administration, it is clear that there have been some successes, including the death of bin Laden and the nuclear arms control treaty with Russia. There have been some notable failures as well, including the precipitous decline in relations with Pakistan. And there are a number of unknowns: What will happen in Afghanistan after 2014 and withdrawal of U.S. combat troops; the changing tone of relations with Russia under Putin (rather than Medvedev); relations with China as it becomes more assertive and dominant; the possible threats posed by a nuclear-enabled Iran and North Korea; ongoing instability in the Middle East; and the lack of resolution of the Israel-Palestine situation? Despite the economic crisis in the Euro-zone, relations with Europe are stable once again, leaving Obama and his national security team free to turn their attention elsewhere.

Understanding present policy, though, requires a brief summary of the past.

Operation Desert Storm, 1990–1991

One of the challenges that the United States had to confront as the Cold War was ending was Iraq's invasion of Kuwait, leading to the first Persian Gulf War in 1991. This crisis emerged relatively quickly and barely pre-dated the implosion of the Soviet Union in August of that year. The military response of the George H.W. Bush administration reinforced the wisdom of the Powell Doctrine regarding when and whether to use force and then, following an affirmative decision, to ensure that the appropriate amount of force is used to achieve the desired objective. As then-Chairman of the Joint Chiefs of Staff, Colin Powell made clear, the importance of the linkage between the use of military force and the desired political objectives.[6] To achieve this, Powell and President Bush structured a military force that they saw as appropriate to achieving those ends.[7]

Writing after the conflict, Powell was prescient in describing why the United States chose *not* to enter Baghdad and pursue Saddam Hussein once the military objectives were achieved: "Even if Hussein had waited for us to enter Baghdad, and even if we had been able to capture him, what purpose would it have served? And would serving that purpose have been worth the inevitable follow-up: major occupation forces in Iraq for years to come and at a very expensive and complex American proconsulship in Baghdad? . . . reasonable people at the time thought not. They still do."[8] That lesson was later lost, leading the United States to confront one of the major challenges it continues to face in Iraq.

The Clinton Administration

The proliferation of ethnic and civil conflicts that followed the end of the Cold War engaged the United States in parts of the world—such as the Balkans and Somalia—that had been seen as remote or outside its "national interest." Despite the fact that they were initiated under the previous administration, it was the Clinton administration that had to follow through and bore the brunt of the criticism for the mistakes that were made.[9] But both of those cases, as well as Haiti in 1993, which was within the traditional sphere of U.S. influence, showed the United States as ill-prepared to deal with the changing face of security and the concomitant military missions that it would have to fight.

The priorities of the Clinton administration were defined partly by domestic politics that included a hostile Republican Congress and a faltering economy as well as foreign policy crises. Having had to address the challenges, and even failures, of his foreign policy, late in his administration Clinton unveiled another "doctrine" regarding the use of force. The *National Security Strategy Report* of 1998 identified three core objectives for U.S. national security strategy that were to be "attuned to the realities of our new era . . . " The three were "to enhance our security, to bolster America's economic security, [and] to promote democracy abroad."[10] But the document also makes clear that security in the 21st century must depend on a number of things: a well-equipped military force that will be ready to fight if and/or when needed on a "wide range of initiatives"; an expanded alliance structure; free trade promoted globally; strong arms control regimes; multinational coalitions to fight international problems such as terrorism and drug trafficking; and "binding international commitments to protect the environment and safeguard human rights."[11] It is the latter clause that has been associated most clearly with the Clinton Doctrine and that was used to justify U.S. involvement in Kosovo in 1999, which in turn accelerated the precedent of using U.S. and NATO forces in out-of-area missions.

There are two other important security policies that Clinton initiated that resonate today. The first was the decision to enlarge NATO. Recognizing the importance of this alliance as well as the need to acknowledge the democracy movements sweeping Eastern Europe, NATO leaders meeting in Brussels in January 1994 reaffirmed the notion that membership should be open to new members, a position advocated by President Clinton. In the final communiqué an invitation was issued by NATO for countries to join a new Partnership for Peace (PfP) program as the next step toward NATO enlargement. This decision was not without controversy both within the United States and also among Alliance members. Where some of the NATO allies could accept the idea of the PfP, they were skittish about formally enlarging the Alliance. In fact, since NATO enlargement was one of the planks in the 1994 Republican Party's Contract with America, they saw the United States as moving ahead unnecessarily quickly for domestic political reasons rather than for reasons of enhancing NATO's military and security roles.

In many ways, the enlargement of NATO and the concomitant politics surrounding that decision changed alliance dynamics, which had implications for the United States later on. For example, when the United States went to war with Iraq in 2003 without UN approval and over the objections of traditional allies France and Germany, it was new allies, including Poland, Albania, Georgia, and Slovakia, that stood with the United States and sent troops, albeit in relatively small numbers.[12]

The second component of Clinton's policy, which was mentioned only in passing at that time, was the need for international commitments to protect against terrorism. Clinton was excoriated for his modest military response to terrorist bombings of U.S. embassies in Kenya and Tanzania in August 1998. In a speech before the UN General Assembly in September 1998, he warned that terrorism is not only an American problem but "a clear and present danger to tolerant and open societies and innocent people everywhere" that "should be at the top of the world's agenda."[13]

U.S. Policy from 9/11

The initial phase of the George W. Bush administration prior to the attacks of 9/11 was generally characterized by drift in the areas of foreign and security policy as the Bush team grappled with questions about how to define U.S. "national interest" in the absence of a Soviet threat, or any other single threat. Writing before the election, Condoleezza Rice, later U.S. National Security Advisor and subsequently Secretary of State, noted that the priorities of the Bush administration would be to pursue its national interests, defined "by a desire to foster the spread of freedom, prosperity and peace."[14] In this statement we see the root of the Bush administration's justification for expanding the use of U.S. military force to spread freedom and democracy.

When the United States made the decision to invade Iraq in 2003 without UN Security Council approval, the virtually unilateral action ultimately was justified based on the need to eliminate a dictator and bring freedom and democracy to Iraq.[15] Even the name of the operation, *Iraqi Freedom*, was fraught with ideology. And it was this ideological perspective that became the justification for much of the use of force during the eight years of the Bush administration.

In many ways, the decision to invade Iraq in March 2003 was the culmination of planning that predated 9/11. "Iraq became the first test case in the Bush administration's new foreign policy doctrine of America's right not only to preeminence in world affairs but to *preemption*, by military might if necessary, of whatever threats it perceives to its security at home and abroad" (emphasis added).[16] The creation and implementation of the Bush Doctrine saw a significant change in the role of the military as an instrument of U.S. foreign—as well as security—policy. But the question that remained was whether the United States had the military forces necessary to support this doctrine, a point that remains under discussion.

By the time that the Bush administration made the decision to go into Iraq, it was already engaged in Afghanistan, a mission that was justified and greeted positively on the whole. Supported initially by the British in what then Prime Minister Tony Blair described as "acting in Britain's national interest,"[17] this ultimately grew to become a NATO mission and embodied the new international role for NATO.

The subsequent invasion of Iraq made clear the weaknesses of the U.S. plans to create a military force capable of fighting two regional wars as outlined by the first President Bush. But also apparent were the flaws in U.S. military planning and understanding of the changing nature of military operations. As noted by H. R. McMaster, "During the decade prior to the terrorist attacks against the United States in September 2001, thinking about defense was driven by a fantastical theory about the character of future war rather than by clear visions of emerging threats to national security in the context of history and contemporary conflict . . . Self-delusion about the character of future conflict weakened U.S. efforts in Afghanistan and Iraq as war plans and decisions based on flawed visions of war confronted reality."[18] Thus, his argument is that the United States took the wrong lessons from the first Persian Gulf War, assuming that the U.S. advantage that they already enjoyed would ensure that "there would be no peer competitor with U.S. military forces until at least 2020."[19] And while military technologies did change the character of war, especially in regard to surveillance and precision munitions, the reality in both Afghanistan and Iraq was that the wars were (and are) largely fought by troops on the ground. Furthermore, success requires more than technological superiority in the types of counterinsurgency that the United States and its allies have had to confront in fighting these wars; they also require the ability to know and understand the culture as well as economic and diplomatic efforts to work with the people if true stability is to be achieved.

The events of 9/11 resulted in the implementation of new policies, boldly packaged in what became known as the Bush Doctrine. For the first time, the United States made it clear that its military policy would be tied to the idea of preemptive attack against any enemy who not only threatened them or its allies but that had to the *potential* to do so. The 2002 *National Security Strategy of the United States*, which codified this policy, outlined clearly the direction for U.S. foreign and military policy for the Bush administration when it proclaimed that "While the United States will constantly strive to enlist the support of the international community, *we will not hesitate to act alone, if necessary, to exercise our right of self-defense by acting preemptively*" (emphasis added). And it also expands the rationale for the use of U.S. military force in stating that "the United States will use this moment of opportunity *to extend the benefits of freedom across the globe*" (emphasis added).[20] Though the United States was already engaged in the war in Afghanistan, these ideas became part of the justification for the subsequent invasion of Iraq. Both of these wars continued well into the Obama administration.

Speaking at West Point in December 2008, Bush again reiterated that the United States was engaged in an ideological struggle, and in order to stay "a step ahead of our enemies," we "transformed our military both to prevail on the battlefields of today and to meet the threats of tomorrow."[21] This transformation included equipping troops with improved real-time intelligence capabilities, increasing the arsenal of unmanned drones, expanding U.S. special operations forces, and placing a new focus on counterinsurgency. It also involved moving troops from Europe, which was seen as part of Cold War thinking, to Asia, and establishing new military commands, including one in Africa, a clear projection of U.S. power globally.[22] But what Bush did not address were the interrelated problems of how to pay for these changes to the military, which is clearly a domestic issue, how these changes in perspective would be received outside the country, and also how to prepare for the peace in Iraq and Afghanistan, all of which are broad security challenges.

The Obama Administration and Afghanistan

Obama has had to face the tasks of rebuilding the U.S. image internationally[23] as well as repairing relations with allies and determining security policy regarding real, or potential, threats while also ending two wars. These are not small tasks and have to be undertaken while also having to address a deteriorating economic situation at home. In spite of their apparent ideological differences the reality is that, at least initially, President Obama continued along the security trajectory defined by his predecessor.

The course of the war in Iraq was set to a large extent by the Status of Forces Agreement signed by the Bush administration in November 2008 with the commitment that U.S. combat forces withdraw by mid-2009. Although this does not suggest the end of U.S. involvement in Iraq, the main role of the military forces shifted from combat to support. Afghanistan proved to be a trickier issue, however.

In December 2009, President Obama announced his decision to send 30,000 troops into Afghanistan over the subsequent six months with plans to pull them out one year later. The belief was that this "surge" would change the momentum and allow the Afghans to take up the fight. The decision was the result of a three-month internal review that was sparked at least in part by a report by General Stanley McChrystal, the U.S. and NATO commander in Afghanistan, offering his assessment of the situation. It concluded that "Failure to gain the initiative and reverse insurgent momentum in the near-term (next twelve months)—while Afghan security capacity matures—risks an outcome where defeating the insurgency is no longer possible," and that the conflict "will likely result in failure."[24] This would require additional resources, but the danger of inaction "risks a longer conflict, greater casualties, higher overall costs, and ultimately, a critical loss of political support."[25] The consequence would be mission failure.

Although the Obama administration review dealt primarily with the number of troops that would need to be deployed to be successful in Afghanistan, costs were also a factor, especially at a time of rising deficits and economic downturn. The Congressional Research Service estimated that the cost of the war in Afghanistan for the 2010 fiscal year was $93.8 billion, up from $59.5 billion the previous fiscal year, and that Iraq had dropped to $71.3 billion in the 2010 fiscal year, from $95.5 billion the year before.[26]

In June 2011, President Obama announced plans to withdraw 10,000 troops from Afghanistan by the end of the year, with 20,000 more to leave by the end of 2012. He said that the reductions of the remaining troops (about 60,000) would continue "at a steady pace . . . bringing to an end America's longest war—a conflict that has cost 1,500 American lives."[27] Most American forces are expected to be out of Afghanistan by 2014. The withdrawals announced "are both deeper and faster than the recommendations made by Mr. Obama's military commanders," and they come amidst ongoing economic issues, as well as a public that has grown weary of war.[28]

Obama's National Security Strategy

In May 2010, eight years after President Bush outlined the Bush Doctrine at West Point, President Obama used the same venue to outline his own security strategy. He made clear the linkage between economic strength and global leadership, noting that although U.S. security depends on a strong military, that is only one component to be complemented by diplomatic initiatives, development experts, and intelligence and law enforcement, all of whom can work to strengthen other countries as well as U.S. national security interests.[29]

President Obama used his speech to introduce the major themes of his security strategy that is "rooted in diplomatic engagement and international alliances," in many ways repudiating the unilateral policies and right to fight a pre-emptive war advocated by his predecessor.[30] He stated clearly that military strength is tied directly to economic strength, and he stressed the importance of strengthening alliances, themes that are stressed in the National Security Strategy document of 2010. In contrast to the National Security Strategy document proposed by President Bush, which is outwardly focused, in 2010 there is a section under "Advancing Our Interests" that pertains specifically to domestic issues, including enhancing science, technology, and innovation, and accelerating sustainable development. The section entitled "Prosperity" states explicitly that "The Foundation of American leadership must be a prosperous economy. And a growing and open global economy serves as a source of opportunity for the American people and a source of strength for the United States."[31] This again underscores some of the ways in which the Obama administration understands the linkage between domestic economic issues and security, defined broadly.

The theme of rebuilding and strengthening ties with allies, both new and older, was a theme that then-candidate Obama stressed, and it was one to which he returned many times. In some ways, part of the challenge facing the Obama administration, given its emphasis on multilateral relationships, was how to rebuild the ties that had been strained or even severed during the eight years of the Bush administration? Obama started to strengthen ties with the traditional European allies as early as July 2008, when as a candidate he traveled to France, Britain, and Germany, where he delivered a speech at the Brandenburg Gate. He made his first official trip to Europe as president in April 2009, three months after taking office. By July 2009, a Pew Global Attitudes Poll reported that "the image of the United States has improved markedly in most parts of the world . . . Improvements in the U.S. image have been most pronounced in Western Europe, where favorable ratings for both the nation and the American people have soared."[32]

President Obama made similar efforts to reach out to other areas, especially the Muslim world, which was important not only to mend relations that had been frayed under President Bush but also because the United States needed to establish credibility globally if it was to regain—or keep—its leadership position. He initiated that process first with a speech to the Turkish Parliament in Ankara (the only Muslim NATO member) in April 2009, followed by a speech at Cairo University in Egypt in June 2009. In that speech, he acknowledged the need for "a new beginning between the United States and Muslims around the world: one based upon mutual interest and mutual respect; and one based upon the truth that American and Islam are not exclusive and need not be in competition. Instead they overlap, and share common principles—principles of justice and progress; tolerance and the dignity of all human beings."[33]

He also sought to "reset" the relationship with Russia, which clearly remains an important strategic partner, if often a difficult one. In April 2010, following a number of meetings, President Obama and Russian President Medvedev signed a nuclear arms pact that would result in deep cuts in the two countries' respective nuclear arsenals, one of Obama's priorities. But the "New START" agreement was also seen as an important step in cementing positive relations between the two countries. What happens now is as much a function of Russian President Putin as well as President Obama.

The result of these initiatives have borne fruit as seen in the largely favorable image of the US globally as measured by the 2011 Pew Global Attitudes Poll. The United States received high marks in Western Europe, "where at least six-in-ten in France, Spain, Germany and Britain rate the U.S. positively." Even in France, where favorability toward the United States was as low as 42 percent and 43 percent in 2003 and 2005, respectively (at the start and escalation of the war with Iraq), it had risen to 75 percent in the most recent poll."[3] Thus, the Obama administration has worked actively to foster ties with other countries, an initiative that is seen as important to enhancing U.S. security.

Obama and Nuclear Weapons

In April 2010, just about the same time the New START Agreement was signed, the U.S. Defense Department issued its *Nuclear Posture Review Report* (NPRR). The document, according to then Secretary of Defense Robert Gates, "defines a roadmap for implementing President Obama's agenda for reducing nuclear risks to the United States, our allies and partners, and the international community." It also notes realistically that "as long as nuclear weapons exist the United States must sustain a safe, secure, and effective nuclear arsenal—to maintain strategic stability with other major nuclear powers, deter potential adversaries, and reassure our allies and partners of our security commitments to them."[35] In short, this document outlines what the United States perceives as the threats from nuclear terrorism and nuclear proliferation and how the these could be addressed, while at the same time ensuring strategic stability with existing nuclear powers, notably China and Russia.

But one can ask what changes the *Nuclear Posture Review Report* really makes regarding U.S. nuclear policy, and the answer is not much. Rather, as Peter Feaver observes, "what may be most important about the new doctrine is the light it shines on the assumptions and strategic logic that motivate the national security thinking of the Obama administration . . . The administration clearly believes that announcing new limits on our nuclear posture will be a strong reason for rogue states to become compliant. This seems hopelessly idealistic: we've given Iran and North Korea plenty of strong incentives with no progress." Feaver concludes that the loopholes built into the document "seem a tacit acknowledgment that [the administration] understands that such idealism is not a reliable guarantor of American national security."[36]

There is little doubt that President Obama is committed to a world in which the use of nuclear weapons is minimized as much as possible, and the United States will pursue policies to help it achieve that goal: maintain a credible nuclear deterrent, modernize aging facilities and weapon systems, promote "strategic stability" with Russia and China, and work to reduce the salience of nuclear weapons globally.

National Security Challenges

The United States is confronting a number of foreign policy/national security challenges that it will have to address. Some of these, such as the rise of China, were predicted, although that does not make them any easier to deal with. Others, such as the revolutions that swept the Middle East, were not. Despite the fact that President George W. Bush advocated for democratic revolutions to sweep the region as the logical follow-up to the end of the Saddam Hussein regime in Iraq, when they started in Tunisia in December 2010 and then spread, the United States had no established policy to deal with them. Coming on the heels of two wars—Afghanistan and Iraq—U.S. military support for the mission in Libya served to underscore how thinly stretched U.S. forces are at this time. Even though plans are in place for U.S. withdrawal from both Afghanistan and Iraq, the reality is that more than ten years of war have taken their toll on the U.S. military, as well as the domestic economy.

The United States is also facing new functional challenges to its security; paramount among these is the threat posed by cyberterrorism. Ironically, the

United States helped raise this danger to the top of the international system when it participated in the development of the Stuxnet virus, which it used against Iran. Now cyber-terrorism is a potential U.S. threat as well.

The Rise of China

Relations with China seem to be especially problematic. As the *Nuclear Posture Review Report* notes, "the United States and China's Asian neighbors remain concerned about China's current modernization efforts, including its qualitative and quantitative modernization of its nuclear arsenal." Moreover, "the lack of transparency surrounding its nuclear programs—their pace and scope, as well as the strategy and doctrine that guides them—raises questions about China's future intentions."[37] It is that uncertainty that is so problematic not only for the United States but also for other states in the region that wonder about China's intentions.

China's rise as a major regional, if not global, power is perhaps one of the major challenges to U.S. security policy for any number of reasons. China is highly integrated into the global economy in general and the U.S. economy in particular, which, to a large extent, constrains the U.S. policy options available. On the one hand, a sound argument can be made that China's integration into and role in the world economy suggest that it is unlikely to engage in any armed conflict that would disturb that balance. On the other hand, though, China's recent actions in Asia, and especially in the South China Sea, have resulted in tensions between China and the United States and its allies in the region, including Japan and the Philippines. This leads to the question of whether China's continued ascendancy can remain peaceful.

Mearsheimer recently elaborated on China's role and what that might mean for the United States and other countries. A quintessential realist, Mearsheimer looks at power balances and claims that "States that achieve regional hegemony have a further aim: they seek to prevent great powers in other geographical regions from duplicating their feat. A regional hegemon, in other words, does not want peer competitors."[38] Assuming that great powers want to dominate at least regionally, Mearsheimer concludes that "China will try to dominate the Asia-Pacific region much as the United States dominates the Western Hemisphere."[39] To Mearsheimer, the result will be the creation of a new "balance of power" regime in the Asia-Pacific region, where China's neighbors will join "an American-led balancing coalition designed to check China's rise."[40] In fact, he can envision a situation in which the U.S. presence in the region will increase, although it will no longer be the preponderant power, having to share that role with China. We can already see that taking place to some extent, as the United States has made it clear that it, too, is a Pacific power.

Joseph Nye also looks at the rise of China's power but introduces yet another set of factors that need to be considered in looking at the relationship between the two countries. Nye, in fact, reminds us that "China has a long way to go to equal the power resources of the United States and it still faces many obstacles to its development."[41] Despite its rapid economic growth, the reality is that China has a vast underdeveloped countryside, an economy that still is not likely to equal the U.S. GDP for decades, an increasing demand for greater political openness, and military capabilities that lag behind the United States. Nye sees the likelihood of the creation of a countervailing coalition anchored by the United States to balance China's powers, but he also notes that "The rise of Chinese power in Asia is contested by both India and Japan (as well as other states), and that provides a major power advantage to the United States."[42] In other words, it is not up to the United States *alone* to provide a balance to China's rising position in the region. At a time when the Obama administration has been advocating for alliance and multilateral action, the situation in the Asia-Pacific region provides an excellent opportunity to put that policy into practice.

Middle East

There are a number of challenges that the United States is facing regarding the Middle East, beyond winding down the existing conflicts. The rapid spread of revolutions that started with Tunisia, resulting in the ousting of President Zine el-Abidine Ben Ali in January 2011, and then spread to Egypt were, in some ways, the result of U.S. encouragement for more democracy in the region. Yet, the United States was unprepared for the uncertainty and instability that followed. Not only does this raise questions about U.S. policy in the region, but it also poses a threat to U.S. ally Israel and lessens still further the likelihood of agreement between Israel and the Palestinian people.

President Bush allowed the Palestinian-Israeli peace process to languish for a host of reasons; in contrast, President Obama indicated that he wanted to make it a priority and stressed that by appointing George Mitchell, the architect of the Good Friday Agreement in Northern Ireland, to serve as his special envoy to the region. After two years of shuttle diplomacy, little progress was made. Mitchell resigned his post in May 2011, effective the same day that Obama was to meet with Israeli Prime Minister Netanyahu. Although that meeting took place, it was clear that little progress was made on the major issues of the 1967 borders and Palestinian statehood.

In a subsequent speech, President Obama outlined his vision in which Israeli-Palestinian relations were put into the larger context of the political changes under way throughout the region. He asked rhetorically "what role America will play as this story unfolds. For decades, the United States has pursued a set of core interests in the region: countering terrorism and stopping the spread of nuclear weapons; securing the free flow of commerce and safe-guarding the security of the region; standing up for Israel's security and pursuing Arab-Israeli peace." He then outlined what U.S. policy would be "to promote reform across the region, and to support transitions to democracy. That effort begins in Egypt and Tunisia, where the stakes are high. . . . Both nations can set a strong example through free and fair elections, a vibrant civil society, accountable and effective democratic institutions, and responsible regional leadership. But our support must also extend to nations where transitions have yet to take place."[43]

The uprising in Libya began in February 2011, and in March, the UN Security Council adopted resolution 1973, authorizing a no-fly zone over the country along with "all necessary measures" needed to protect civilians. A coalition of nations, including the United States, ultimately was involved with a bombing attack designed to aid rebel anti-Qaddafi forces. In October 2011, the rebel group was victorious, and ultimately Qaddafi was found and killed. Currently, there is a transitional government in place, but as the attack on the U.S. consulate in Benghazi illustrates clearly, the situation is far from stable.

The situation in Libya, especially the decision to support the rebel forces against Qaddafi, stands in marked contrast to U.S. policy regarding the civil war in Syria, which presents a far greater issue for the United States. On the one hand, civil war has raged for more than two years, with no end in sight. In a defiant speech on January 6, 2013, President Assad gave no indication that he plans to step down nor are there any indications that the violence will abate. He also ruled out any possibility of negotiating with the opposition. According to UN estimates, more than 60,000 have died in the civil war since the uprising started, as what was then a peaceful protest grew into an armed struggle.[44] Determining how to handle this ongoing situation has posed a dilemma for the Obama administration.

In the third presidential debate, questions were raised about U.S. policy toward the Middle East in general and Syria in particular, especially the contrasts between the ways in which the United States responded to the uprising in Libya versus the country's policy regarding Syria. The president was blunt when he said "When we went into Libya, and we were able to immediately stop the massacre there, because of the unique circumstances and the coalition that we had helped to organize." But in perhaps the clearest statement of his administration's policy, especially in contrast to the Bush administration, President Obama stated clearly that "Syrians are going to have to determine their own future." And he made clear the often-conflicting demands that come with "nation-building" when he said "You know, one of the challenges over the last decade is we've done experiments in nation building in places like Iraq and Afghanistan and we've neglected, for example, developing our own economy, our own energy sectors, our own education system. And it's very hard for us to project leadership around the world when we're not doing what we need to do." [45]

As we look to the challenges of a still unsettled situation in parts of the Middle East, it appears that U.S. policy will be to intervene when necessary as long as the objectives are clear but will not engage in nation-building, especially when it comes at the expense of domestic needs and priorities.

Cyber-terrorism

In remarks at the White House in May 2009, President Obama noted that "It's long been said that the revolutions in communications and information technology have given birth to a virtual world . . . This world—cyberspace—is a world that we depend on every single day . . . It's the classified military and intelligence networks that keep us safe, and the World Wide Web that has made us more interconnected than at any time in human history." And he then raised the following point: "*So cyberspace is real. And so are the risks that come with it*" (emphasis added).[46]

Those risks have been borne out in a number of circumstances and settings. Early in January 2013, it was reported that hackers had attacked the online sites of several American banks. These seemed to be the second stage of a campaign, started in September 2012, against America's largest banks, including Bank of America, Citigroup, and Wells Fargo. The alleged hackers claimed that "it had attacked the banks in retaliation for an anti-Islam video that mocked the Prophet Muhammad and pledged to continue its campaign until the video was removed from the internet."[47] Of the nine banks allegedly attacked, representatives of three confirmed disruptions to their on-line banking sites. Security researchers who have studied these attacks confirmed that the hackers had used a new type of weapon to disrupt the banking. Security researchers "still do not know how the data centers used in the first wave of attacks were infected in the first place, how widespread the infection rate was and—perhaps most troubling—whether the servers could be used to damage other sensitive targets in the future." One thing that was clear was that these attacks were not sponsored by a country.[48]

Although private individuals with technological know-how—and potentially terrorist groups—clearly can disrupt major institutions such as banks, we also know that there have been various cyber-attacks perpetrated by countries, which suggests a new threat and potentially a different form of warfare. For example, China penetrated Google and more than thirty other companies in January 2010.[49] And the United States, allegedly working with Israel, was involved with perpetrating an attack against Iran's nuclear facilities using the Stuxnet virus, which according to some reports pushed Iran's nuclear program back about eighteen months. In the latest example of cyber-escalation, Iran has posted a message on its web site indicating that it was developing its own virus that was "potentially more harmful than Stuxnet."[50]

In October 2012, then Secretary of Defense Leon Panetta warned that the United States "was facing the possibility of a 'cyber-Pearl Harbor' and was increasingly vulnerable to foreign hackers who could dismantle the nation's power grid, transportation system, financial networks and government."[51] Panetta "painted a dire picture of how such an attack on the United States might unfold . . . he was reacting to increasing aggressiveness and technological advances by the nation's adversaries, which officials identified as China, Russia, Iran and militant groups . . . [and] the most destructive possibilities . . . involve cyber actors launching several attacks on our critical infrastructure at one time in combination with a physical attack."[52]

President Obama raised a number of other important points that are directly relevant when he noted that "Our technological advantage is a key to America's military dominance. But our defense and military networks are under constant attack . . . it's now clear this cyber threat is one of the most serious economic and national security challenges we face as a nation. It's also clear that we are not as prepared as we should be . . . just as we failed in the past to invest in our physical infrastructure—our roads, our bridges and rails—we've failed to invest in the security of our digital infrastructure."[53]

The *National Security Strategy* devotes a major section to the topic of "Secure Cyberspace," which outlines strategies to address and protect the United States from the threat posed by various aspects of cyber-attacks. The document makes it clear that "Our digital infrastructure . . . is a strategic national asset, and protecting it . . . is a national security priority."[54] The threat from a cyber-attack is a very real one that has emerged to be a high security priority.

Reflections

The debate over raising the debt ceiling in summer 2011 and the "fiscal cliff" debate of December 2012 illustrate clearly the relationship between economic stability and national security. It is incumbent upon the United States to adjust its thinking about what "security" means and what is in its national interest. To quote Leslie Gelb, "Washington still principally thinks of its security in traditional terms and response to threats with military means. The main challenge for Washington, then, is to recompose its foreign policy with an *economic theme*, while countering threats in new and creative ways. The goal is to redefine 'security' to harmonize with twenty-first century realities" (emphasis added).[55] Gelb also criticizes the United States for being late in adjusting its policies to address this global change in the importance of economics, despite recognizing it.

Clearly, there are reasons for this tardiness, relative to other countries that have adjusted far more quickly. "For over half a century, U.S. foreign policy had to give priority to stemming real and serious threats . . . The United States cannot ignore these burdens, but it must adjust its approach to recognize that *economics* is now at the center of geopolitics. Washington's failure to do so has already cost it in blood, treasure and influence" (emphasis added). But, even more important, Gelb states, "leaders in Washington proclaim their awareness of the new economic order but lack any semblance of a new national security strategy to embrace it."[56]

Gelb's point is that the major powers again need one another if their economies are to grow. Conversely, as we have seen, wars sap economies, and major powers have more to lose from engaging in such wars than they have to gain. President Obama seemed to recognize the importance of economics, as indicated in the *National Security Strategy* of May 2010. But recognizing this and doing something about it, especially in the face of a hostile Republican Congress, is not the same thing.

In an "Opinion" piece in *The Wall Street Journal* in July 2011, Walter Russell Mead noted that despite the dire predictions about the decline of the United States, realistically, the country is better positioned to withstand the challenges of the 21st century than any other country. In this piece, he argued that rather than look at current international issues such as the rise of China as a challenge, they can be viewed as a change that can support U.S. national interests in Asia. "Unlike Europe in 1910, Asia today looks like an emerging multipolar region that no single country, however large and dynamic, can hope to control. This fits American interests precisely. The U.S. has no interest in controlling Asia or in blocking economic prosperity that will benefit the entire Pacific basin, including our part of it."[57]

The point here is that the United States needs to stop looking at the rise of China as coming *at the expense of* the United States. Rather, at a time when U.S. and China economies are co-mingled, the two should be seen as strategic partners, if not allies. Furthermore, as Mead suggests, the rise of China might require that the United States rethink its policies regarding Asia and the Pacific, resulting in a more proactive policy that supports what is in the U.S. national interest, rather then putting the country in a position of having to react to the policies of other countries.

But this point also underscores the importance of economic security as a component of national security writ large. In the discussions and debates surrounding the need to raise the debt ceiling, one of the few areas that all sides could agree on was the need to cut military spending beyond the savings that would be accrued with the winding down and subsequent ending of the wars in Afghanistan and Iraq. This suggests that the United States could be secure militarily even with cuts in what has been seen as a bloated Defense Department. But any cuts, when they come, must be tied back to the basic question of what the United States needs in order to ensure its own security and that of its allies. Is security a function of more weapons, or is it really a function of having a force, both personnel and equipment, that can meet the anticipated needs?

Clearly, no leader(s) can ever anticipate every possible contingency; few could have predicted the fall of the Soviet Union or the revolutions of the

Middle East. But any student of foreign and security policy does know and understands that the needs of the future will be very different from those of the past. The question facing the United States is how well learned are the lessons of the past, and what policies can best ensure a secure future?

Notes

1. Robert Kagan, *The World America Made* (New York: Alfred A. Knopf, 2013, 8–9).

2. See Martin S. Indyk, Kenneth G. Lieberthal, and Michael E. O'Hanlon, *Bending History: Barack Obama's Foreign Policy* (Washington, D.C.: The Brookings Institution, 2012).

3. See David E. Sanger, *The Inheritance: The World Obama Confronts and the Challenges to American Power* (New York; Three Rivers Press, 2009). Also see Sanger, *Confront and Conceal: Obama's Secret Wars and Surprising Use of American Power* (New York: Crown Publishers, 2012).

4. See "National Security Strategy," May 2010, 27, http://www.whitehouse.gov/sites/default/files/rss_viewer/national_security_strategy.pdf. See also Sanger, *Confront and Conceal*, op. cit.

5. At his speech to the Australian Parliament, November 17, 2011, President Obama stated, "After a decade in which we fought two wars that cost us dearly, in blood and treasure, the United States is turning our attention to the vast potential of the Asia Pacific region . . . Our new focus on this region reflects a fundamental truth—the United States has been, and always will be, a Pacific nation." http://www.whitehouse.gov/the-press-office/2011/11/17/remarks-president-obama-australian-parliament.

6. See Colin L. Powell, "U.S. Forces: Challenges Ahead," *Foreign Affairs* (Winter 1992/93) http://www.cfr.org/world/us-forces-challenges-ahead/p7508.

7. Richard N. Haass, *War of Necessity, War of Choice: A Memoir of Two Iraq Wars* (New York: Simon & Schuster, 2009), 98–99.

8. See Colin L. Powell, "U.S. Forces: Challenges Ahead," *Foreign Affairs* (Winter 1992/93) http://www.cfr.org/world/us-forces-challenges-ahead/p7508.

9. Elizabeth Drew, *On the Edge: The Clinton Presidency* (New York: Simon & Schuster 1994), 323.

10. The White House, "A National Security Strategy for a New Century," October 1998, http://www.globalsecurity.org/military/library/policy/national/nss-9810.pdf.

11. Ibid., "Preface."

12. According to Todd S. Purdum, of the 21,000 non-American troops in Iraq in August 2003, about four months after the invasion, 11,000 were British and the other nations including "the other 'coalition' partners—nations like Albania, Estonia, Georgia, Latvia, Lithuania, Poland, Slovakia and Spain—had each contributed an average of fewer than 600 troops. The high cost of go-it-aloneism was becoming apparent. . . . "

Todd S. Purdum, *A Time of Our Choosing: America's War in Iraq* (New York: Times Books, 2003), 264.

13. "Remarks of President Clinton to the UN General Assembly," September 21, 1998, http://daccessdds.un.org/doc/UNDOC/GEN/N98/858/28/PDF/N9885828.pdf?openElement.

14. Condoleezza Rice, "Promoting the National Interest," *Foreign Affairs*, vol. 79, no. 1 (Jan.–Feb. 2000), p. 62.

15. George W. Bush, "The Struggle for Democracy in Iraq: Speech to the World Affairs Council of Philadelphia, December 12, 2005, http://www.presidentialrhetoric.com/speeches/12.12.05.html. See also Philip H. Gordon, *Winning the Right War: The Path to Security for America and the World* (New York: Times Books, 2007), and Purdum, *A Time of Our Choosing, op. cit.*

16. Purdum, *A Time of Our Choosing*, 4.

17. Joyce P. Kaufman, "The United States and the Transatlantic Relationship: A Test for U.S. Foreign and National Security Policy," in *The Future of Transatlantic Relations: Perceptions, Policy and Practice*, Andrew M. Dorman and Joyce P. Kaufman, eds. (Stanford, Calif.: Stanford Security Series, 2011), p. 65.

18. H. R. McMaster, "On War: Lessons to Be Learned," *Survival*, vol. 50, no. 1 (February–March 2008), 19.

19. Ibid., 20.

20. "The National Security Strategy of the United States of America," September 2002, http://georgewbush-whitehouse.archives.gov/nsc/nss/2002/print/index.html.

21. "President Bush Discusses Defense Transformation at West Point, December 9, 2008, http://georgewbush-whitehouse.archives.gov/news/releases/2008/12/print/20081209–3.html.

22. Ibid.

23. A 2008 Pew poll found that "the U.S. image is suffering almost everywhere," due, at least in part, to the fact that "people blame America for the financial crisis." Pew Global Attitudes Project, "Global Opinion in the Bush Years (2001–2008)," http://global/org/reports/pdf/263.pdf, 1.

24. Quoted in Bob Woodward, "McChrystal: More Forces or 'Mission Failure,'" *The Washington Post*, September 21, 2009, http://www.washingtonpost.com/wp-dyn/content/article/2009/09/20/AR2009092002920.html.

25. Quoted in Woodward, "McChrystal: More Forces or 'Mission Failure.'"

26. Amy Belasco, "The Cost of Iraq, Afghanistan, and Other Global War on Terror Operations Since 9/11," Congressional Research Service Report for Congress, March 29, 2011, http://www.fas.org/sgp/crs/natsec/RL33110.pdf.

27. Mark Landler and Helene Cooper, "President Orders Rapid Troop Cuts in Afghanistan," *The New York Times*, June 23, 2011, p. A11.

28. Ibid.

29. Text of Obama's Speech to West Point 2010 Cadets, May 22, 2010, http://www.cbsnews.com/stories/2010/05/22/national/main6509577.shtml.

30. Peter Baker, "President Outlines National Security Strategy Rooted in Diplomacy and Alliances," *The New York Times,* May 23, 2010, A16.

31. "National Security Strategy," May 2010, http://www.whitehouse.gov/sites/default/files/rss_viewer/national_security_strategy.pdf.

32. Pew Global Attitudes Project, "Confidence in Obama Lifts U.S. Image Around World," July 23, 2009, www.pewglobal.org.

33. "Remarks by the President on a New Beginning," Cairo University, Cairo, Egypt, June 4, 2009, http://www.whitehouse.gov/the_press_office/Remarks-by-the-President-at-Cairo-University-6–04–09/.

34. Pew Global Attitudes Project, "U.S. Favorability Ratings Remain Positive: China Seen Overtaking U.S. as Global Superpower," July 13, 2011, http://pewglobal.org/2011/07/13/chine-seen-as-overtaking-us-as-global-superpower/.

35. Department of Defense, "Nuclear Posture Review Report," April 2010, http://www.defense.gov/npr/docs/2010%20nuclear%20posture%20review%20report.pdf.

36. Peter D. Feaver, "Obama's Nuclear Modesty," Op-ed piece, *The New York Times,* April 9, 2010, A19.

37. "Nuclear Posture Review Report," op. cit.

38. John J. Mearsheimer, "The Gathering Storm: China's Challenge to U.S. Power in Asia," *The Chinese Journal of International Politics*, vol. 3 (2010), 388. Downloaded from http://mearsheimer.uchicago.edu/pdfs/A0056.pdf.

39. Ibid., 389.

40. Ibid., 391.

41. Joseph S. Nye, "The Future of American Power: Dominance and Decline in Perspective," *Foreign Affairs* (Nov/Dec. 2010), 4.

42. Ibid., 5.

43. Text of "Obama's Middle East Speech," *The New York Times*, May 19, 2011, http://www.nytimes.com/2011/05/20/world/middleeast/20prexy-text.html?_r=1&pagewanted=print.

44. See Anne Barnard, "Defiant Speech by Assad is New Block to Peace in Syria," *The New York Times,* January 6, 2013, http://www.nytimes.com/2013/01/07/world/middleeast/syria-war-developments.html?pagewanted=1&ref=world.

45. "2012 Presidential Debate," President Obama and Mitt Romney's remarks at Lynn University on October 22, 2012 (full transcript), *The Washington Post*, http://www.washingtonpost.com/politics/decision2012/2012–presidential-debate-president-obama-and-mitt-romneys-remarks-at-lynn-university-on-oct-22–running-transcript/2012/10/22/be8899d6–1c7a-11e2–9cd5–b55c38388962_story_4.html.

46. Text: "Obama's Remarks on Cyber-Security," May 29, 2009, *The New York Times*, http://www.nytimes.com/2009/05/29/us/politics/29obama.text.html?pagewanted=all (accessed June 4, 2012).

47. Nicole Perlroth, "U.S. Banks Again Hit by Wave of Cyberattacks," *The New York Times*, January 4, 2103, http://bits.blogs.nytimes.com/2013/01/04/u-s-banks-again-hit-by-wave-of-cyberattacks/.

48. Ibid.

49. John Markoff, David E. Sanger, and Thom Shankar, "Cyberwar: In Digital Combat, U.S. Finds No Easy Deterrent," *The New York Times*, January 26, 2010, http://www.nytimes.com/2010/01/26/world/26cyber.html?ref=cyberwar&pagewanted=prin t(accessed June 4, 2102).

50. "Cyberattacks on Iran—Stuxnet and Flame," *The New York Times*, updated August 9, 2012, http://topics.nytimes.com/top/reference/timestopics/subjects/c/computer_malware/stuxnet/index.html.

51. Elizabeth Bumiller and Thom Shanker, "Panetta Warns of Dire Threat of Cyberattack on U.S.," *The New York Times*, October 11, 2012, http://www.nytimes.com/2012/10/12/world/panetta-warns-of-dire-threat-of-cyberattack.html?pagewanted=all&_r=0.

52. Bumiller and Shanker, "Panetta Warns of Dire Threat of Cyberattack on U.S." http://www.nytimes.com/2012/10/12/world/panetta-warns-of-dire-threat-of-cyberattack.html?pagewanted=all&_r=0.

53. Text: "Obama's Remarks on Cyber-Security," May 29, 2009, *The New York Times*, http://www.nytimes.com/2009/05/29/us/politics/29obama.text.html?pagewanted=all (accessed June 4, 2012).

54. "National Security Strategy," May 2010, 27, http://www.whitehouse.gov/sites/default/files/rss_viewer/national_security_strategy.pdf.

55. Leslie H. Gelb, "GDP Now Matters More than Force: A U.S. Foreign Policy for the Age of Economic Power," *Foreign Affairs*, vol. 89, no. 6 (Nov.–Dec. 2010), p. 35.

56. Leslie H. Gelb, "GDP Now Matters More than Force: A U.S. Foreign Policy for the Age of Economic Power," *Foreign Affairs*, vol. 89, no. 6 (Nov.–Dec. 2010), 38.

57. Walter Russell Mead, "The Future Still Belongs to America," *The Wall Street Journal*, July 2–3, 2011, A13.

PART II

EUROPE—THE OLD WORLD

3 France and National Security

Adrian Treacher

Introduction

It could be argued that France was at the apogee of its relative political and military power as Napoleon's armies won victory after victory across the European mainland during the early years of the 19th century and as the Emperor's relatives were placed on the throne of numerous subjugated states. From this narrow, realist perspective, the line of reasoning would continue by claiming that the country has been in steady, relative decline ever since, as it was relegated from a status of elite global superpower to its contemporary one of a medium power among many. This process has encompassed several severe jolts to the national psyche since Napoleon's ultimate demise in 1815, successive humiliating military defeats to the Prussian (1870–1871) and German neighbor (1940), the traumatic loss of empire, and the rise of new superpowers. Parallels could be drawn with the United Kingdom's experiences during the 20th century when its power status was similarly relegated, or with post–Cold War Russia. All three of these countries have not simply accepted this process and settled into marginal roles in world affairs. Rather, through various means, they have each in their own way sought to maximize and even exaggerate their international weight.

Yet such a narrative would present an over-simplified understanding of France today. "A superpower in the seventeenth century, a great power by the time of the First World War, this mature west European country, although now only a 'medium power,' manages to sustain into the twenty-first century an almost unparalleled level of European and global economic and strategic influence given the extent of its demographic and geographic resources."[1] The country has been able to remain at the center of much that is happening around the globe, from its imperial clashes with Britain and other powers during the 18th and 19th centuries (with American independence being just one by-product) to its present day permanent seat at the United Nations Security Council (UNSC), its membership of a still quite restricted club of countries with a viable nuclear weapons capacity, its central position in European integration focused on the European Union (EU), its participation in numerous international organizations, and its leadership of a significant francophone commonwealth.

France and the World: Past, Present, Future

Although not the only indicator of a global presence, the continued prevalence of the French language points to an imperial past on a massive scale encompassing not just Europe but also huge swathes of Africa, the Middle East, North America, and Southeast Asia, as well as numerous other territories across the Caribbean and the Pacific. But France's worldwide influence was not just based on military conquest. French political culture in the 13th and 14th centuries developed a sense of *grandeur*, or national greatness, that claimed continuity from the Roman Empire and all its attributes. Louis XIV would go the furthest in laying claim to this heritage that, it was argued, bestowed the French with special privileges "as the defenders of European civilization itself, not simply of France and its interests."[2]

Following the 1789 revolution, this sense of self-worth was transferred more specifically to the nation as a whole (witness the 1790 declaration of *la grande nation*[3]). The subsequent years, as the country went from The Convention to The Directory and then Imperium would mark further refinement of this self-image with the emergence of France's universal mission based on its values (not least the 1789 Declaration of the Rights of Man and the Citizen, or the role played by Frenchmen who crafted the values of the 18th century enlightenment that were embraced by the 1776 American Declaration of Independence). This claim to France's special role in the world would be unflinching even in the face of the successive national disasters mentioned here; it was eternal, not transitory. "[W]hat had emerged in France by the twentieth century was a political structure with a strong motivation to pass on its special message to the rest of the world, and a belief in the prerogative to speak on behalf of humanity."[4] As James F. McMillan has argued, French leaders had little difficulty in equating the universal cause of freedom and justice with the independence and greatness of their own country.[5] See, for example, the French lead in the 2011 NATO operation over Libya.

This interpretation of the past was a driving force behind the principles incorporated within France's national security thinking and its place in the world as established by General Charles de Gaulle as a primary architect of the new 1958 Constitution and as president of the present Fifth Republic for its first eleven years.[6] Although vague and open to multiple interpretations, *Gaullism* became and remains the guiding rationale for French foreign and defense policy; it has been virtually unchallenged in terms of national debate and across the mainstream political spectrum. Its main tenets, the *grandes lignes*, rest on the aforementioned belief in the country's inherent greatness and on an unwavering belief that France should have an elevated rank and sit among the very top tier of international powers. Central to these Gaullist principles is an emphasis on self-reliance and the notion that France should not be reliant on other powers for its national security. This partly explains de Gaulle's decision to withdraw France from NATO's integrated military structures in 1966.[7] In part, this thinking stemmed from a certain mistrust of the United States as the guardian of the Western world. Certainly there was the perception that American forces had been held back too long in both World Wars (three and two years, respectively) and that the United States had then conspired to marginalize de Gaulle's Free French during the key allied negotiations and summitry vis-a-vis the future design of the global political order toward the end of World War 2. This perception was then given renewed justification, first in 1954 when a tardy U.S. intervention ensured the collapse of France's colonial ambitions in Indo-China and second in the 1956 Suez Crisis, where strong American opposition was a key factor behind the withdrawal of the Franco-British-Israeli forces deployed in Egypt to reverse the nationalization of the Suez Canal.[8] This suspicion of American motives was also transferred, in part at least, to the United Kingdom, which, from a French perspective, became guilty by association. The UK was seen as the willing instrument of American interests and as part of an Anglo-Saxon plot to obtain and secure leadership over the Western world.[9]

Meanwhile, national pride was being swallowed as France's economic recovery rested in no small part on the U.S. Marshall Aid plan. The lesson seemingly being drummed home was that "the failure of alliances (prior to World War 1) and of collective security (prior to World War 2, with the United States remaining outside the League of Nations) meant that France had to henceforth unilaterally secure its own interests without reliance on any external benefactor."[10] As noted by Pierre Lellouche, the defense of France's interests could not be provided, let alone guaranteed, by any outside power and hence France would have to take care of this itself.[11] This lesson, when added to other contemporaneous jolts to the national psyche, notably the crisis in Algeria and the general weakness of the Fourth Republic, led, according to John Gaffney, to a general sense of vulnerability becoming a cultural fact that would impact on the French strategic imagination.[12]

This new emphasis on nondependence (on other actors) and self-reliance translated for example and in notable contrast to its closest and most significant international partner, Germany. It also meant that the France of the Fifth Republic became consistently prepared to use coercive, preventive, military force in pursuit of its perceived national interests and in the name of global security and stability. Invariably, this was on a unilateral basis, notably in sub-Saharan Africa, and sometimes did not even seek validation from the UNSC. However, since the end of the Cold War there has been a perceptible shift of emphasis toward multilateral missions. As noted by Joachim Krause, a primary motivation for this shift has been the desire for greater international legitimacy.[13] This has not resulted in a total abandonment of the unilateral option, as illustrated by this decade's mission in Côte d'Ivoire. Nevertheless, multilateralism is now a central tenet of French global ambitions because it puts emphasis on the primacy of the UNSC, permanent membership of which is a key element of France's continued claim to great power status and a global leadership role. Implicit, and sometimes explicit, in this has been a concern to deter or counter American unilateralism and to assuage concerns about what has been perceived in Paris as Washington's pretensions to the world's hegemon that would marginalize French influence.

This prioritization of the UNSC, and the UN more generally, also goes back to the principles established by de Gaulle. Once he had overcome his suspicions of the organization as an instrument of U.S. foreign policy that interfered in his country's decolonization process, he came to see the opportunities it presented as he sought to reestablish France's position among the world's top powers. Henceforth, French foreign policy has been motivated in part by a concern to ensure that the UN, and particularly the UNSC, acts as *the* primary authority on global issues, especially the maintenance of international peace and stability. France has been able to maximize its global influence and standing so long as the UNSC has not been marginalized. The particularly vicious Franco-American diplomatic crisis of 2002–2003 concerning the invasion of Iraq can be partly understood, from a French perspective at least, in these terms.[14]

The development and retention of France's independent nuclear strike force, the *force de frappe*, has also served the national pursuit of the *grandes lignes*. As the Cold War took hold and the nuclear option sat the apex of weaponry, it was imperative that France be at the forefront of this new mode of warfare and defense, not least to ensure there would be no further national military humiliations. So France's nuclear weapons program was launched during the 1950s during the Fourth Republic and was then prioritized by de Gaulle. Indeed in practice, the *force de frappe* has served a more symbolic and diplomatic role outweighing any strategic value. It has been used to demonstrate and vindicate the country's ongoing pretensions to an elevated rank among states, to serve as a symbol of national independence, and as a means "to mitigate the psychological damage created by relative national decline."[15] It has also been part of another Gaullist principle, namely self-reliance, or, put another way, a policy of political and military nondependence on any other Great Power and of creating as much freedom of maneuver as possible on the international stage. That is not to say that at times of heightened Cold War tension, France did not stand shoulder to shoulder with its Western partners.

Defining Contemporary French National Security

In addition to the continuous stream of declarations by France's political representatives over the decades (primarily the president but also key figures like the defense minister and prime minister) as they have striven to almost instinctively perpetuate the Gaullist principles, analysts of the country's definition of its post–Cold War defense and national security can look at two White Papers (1994[16] and 2008[17]), both of which heralded in subsequent and significant alterations to foreign and security policy. Interestingly, these two represent only the second and third incarnations of a formal strategy paper in this area of policy during the lifetime of the Fifth Republic (the first being in 1972). As Camille Grand observed, it simply has not been a major part of France's military tradition or strategic culture, and as a result there has been and remains no legal requirement for any such document to be produced on a regular basis; that said, 2006 saw the *White Paper on Internal Security and Terrorism* and 2008 the *White Paper on Foreign Policy.*[18]

The 2008 White Paper is understandably a reflection of the views and priorities of President Sarkozy and his administration (in office since May 2007), but, in a break with the previous Papers, there were other inputs too; notably a cross-parliamentary commission that also incorporated high-level representatives from relevant government agencies and the armed forces as well experts from the research community and which held hearings with over fifty personalities from home and abroad and undertook visits in the field. As such, the Paper contains elements of both continuity and change from its 1994 predecessor. For the former, one can point to the continued transfer of national ambitions on to the European Union (EU)[19] and the ongoing staunch commitment to the *force de frappe*, France's nuclear deterrence capability, with France maintaining two separate systems (air-launched cruise missiles and submarine-based ballistic missiles), unlike the UK, which is reliant exclusively on its four ballistic missile-carrying submarines. Both Papers preceded major attempts to redesign France's relations with NATO but only the 2008 version produced tangible changes. Finally, both Papers fed into significant military reform with a focus on reducing the overall size of the military while enhancing its ability to enact force projection overseas (the president can now call upon predominantly professional armed forces comprising roughly 250,000 personnel).[20] With regard to policy formation, meanwhile, Grand reports that overall, the 2008 White Paper reflected general consensus "among French elites and decision-makers that went beyond political lines."[21]

In terms of territorial scope, the 2008 White Paper prioritizes a geographic axis that is spread from the Indian Ocean, across the Persian Gulf, into the Mediterranean and over to the Atlantic. In line with policy formation of previous decades, French national security interests are equated to those of "Europe" (meaning not just the EU but also France's other European partners). Regarding the EU specifically, the French message remains one of enhancing its ability to undertake military and civilian security missions within the European space and across the globe. There has been clear frustration in Paris at the lack of progress in this regard at the EU level. For example, the White Paper calls for an EU military intervention force of 60,000 troops deployable for a year—something that still has not been attained despite this having been agreed at the Helsinki European Council in December 1999.

Through the 1990s, France, along with the UK, had been working to augment security provision across Africa by facilitating crisis management and peacekeeping capabilities from within the continent. The 2008 White Paper places new emphasis on the build-up of Africa's own security instruments as well on the strengthening of cooperation with Europe. And it builds upon initiatives like the Franco-British battle group concept that's now being embraced across Europe, and which was designed with the explicit assumption that it was earmarked for Africa.

What particularly sets the 2008 White Paper apart is the focus on national security that embraces both deliberate threats and accidental or "natural" challenges. Certainly, it acknowledges that there is greater instability and unpredictability in world affairs, and consequently emphasis is placed on enhancing France's intelligence and information gathering capacity so as to obtain a higher level of predictability and preparedness. And once again national security interests are transposed to the European level with the prioritization of combating organized crime and international terrorism (with special emphasis placed on the defense against cyber attack), creating an effective EU civil protection capacity and ensuring security of supply for energy and key strategic raw materials.

The White Paper is quite bullish regarding the defense budget, with spending for 2008–2012 being constant, in line with inflation, and thereafter being a percentage point above inflation. The subsequent global financial and economic crisis threw all that up in the air, and France, like its European neighbors, is now confronted with high national debt levels and pressure to reduce the level of public spending as part of fiscal austerity measures.[22] Inevitably, this is having a knock-on effect for defense spending with the likelihood that France will seek further cross-national cooperation in an attempt to cut costs and that even greater scrutiny will be placed on the merits of military procurement proposals.

The French National Security Apparatus (Internal Instruments, External Partnerships)

With the 2008 White Paper, the range of instruments involved in national security is deliberately extended to the nonmilitary sphere. The armed forces and their agencies were to interact effectively with internal security services and civil security services as part of a renewed emphasis on the protection of the French population and territory. The White Paper stated, "Reinforcing resilience requires a change in the means and methods of surveillance used over the national territory."[23]

In terms of external partnerships, the European Union (EU) steadily became a primary focus for the perpetuation of the *grandes lignes*. A country in the face of a similar set of shocks and reverses to those faced by France (as set out in the previous section) might have quietly, if reluctantly, accepted the new state of affairs and slid off to the margins of the international system; but this was just simply not an option for de Gaulle. So rather than scaling back the scope of France's national security interests and foreign policy objectives to reflect the new realities, he formulated the *grandes lignes* and turned to France's West European partners to ensure their attainment. As mentioned, the UK has likewise resisted this notion of national decline and sought to retain Great Power status, albeit in a different manner. But for France, the EU would enable it to continue to do what it could no longer realistically do alone—remain at the top table of international players. So France's neighbors would be manipulated into a cohesive global political actor under implicit French leadership. Jonathan Eyal noted the remarkable ability of French leaders "to portray their country's national interests as Europe's fundamental imperatives."[24] Post–Cold War, this would be partly illustrated by their obsession with the formation of an autonomous European security and defense[25] capacity (meaning autonomous from the United States).

Reasons behind a general questioning of the automaticity of American support of French interests have been elaborated in the preceding sections. Through much of the 1990s however, French military ambitions for Europe were not realized. The autonomous aspect particularly concerned the two other main European players—Britain and (reunified) Germany—and meant that France's vision had to be amended. These diverging perspectives among France and her key EU partners would be graphically illustrated by their contrasting responses to the collapse of Yugoslavia. Jacques Chirac, as president from 1995, thus accepted that the key debates regarding the management of Europe's security were actually taking place inside NATO and that France needed to reintegrate itself fully with that organization if were to truly be able to influence those debates. Efforts to that effect largely failed, but the negative implications of this for French national interests were significantly softened by the British decision in 1998 to remove objections to an augmented and autonomous European military capacity centered on the EU. Within six months, at Cologne, the Union launched what is now the Common Security and Defence Policy (CSDP), and within just a few years it was undertaking, albeit small-scale, its own military missions, notably in Central Africa and the Western Balkans.[26] That said, the initial momentum behind the CSDP has slowed somewhat over time, and French visions for a fully autonomous European military capacity have been frustrated, not least because, as noted by Ben Jones, key partners like Britain have been diverted by preoccupations with its operations in Iraq and Afghanistan.[27]

As mentioned previously, unlike during the 1990s, recent years have seen "successful" efforts by France to fully rehabilitate itself with the Atlantic Alliance. The 2008 White Paper states that there is no competition between the EU and NATO; rather, it claims, they are complementary. Subsequently, as advocated by the Paper, France has re-entered the integrated military structures of NATO, which it had vacated in 1966. That said, and conforming to the principles set out by de Gaulle, France retains full control of its nuclear and conventional forces, denies any automaticity of military commitment, and retains the right to develop autonomous strategic capabilities (notably in the area of intelligence).[28]

At the bilateral level, the two most significant relationships for France's national security are certainly Germany (especially in terms of economic and political cooperation) and probably Britain (as its key military partner). Following their convergence over the CSDP it was soon apparent that France and Britain shared contrasting visions for the long-term objectives of this new EU policy area in terms of whether it was to serve as an alternative or complement to NATO. This tension has been, partly at least, superseded by France's full reintegration into the Atlantic Alliance. Regardless of political differences, however (another example would be the diplomatic schism concerning the 2003 American-led invasion of Iraq), the two countries have been able to engage in ever deeper military cooperation. This process, alongside other similar bi- and trilateral agreements with other partners, has become even more of a necessity as the financial pressures and constraints of the present global economic crisis bite ever deeper. Whereas previously France had been using its European neighbors to pool political and diplomatic resources in order to continue its pursuit of the *grandes lignes*, there was now the additional imperative of the pooling of military assets. In September 2010, French Defense Minister

Hervé Morin spoke of the need for European countries to counter the need for austerity by focusing on military collaboration or risk being "under a joint dominion of China and America."[29] The following November then witnessed the signing of a fifty-year Franco-British defense and security cooperation treaty that includes the formation of a joint expeditionary force and an agreement on aircraft carrier strike groups; a nuclear weapons cooperation treaty was also signed at this time. Momentum was then maintained by the February 2012 Franco-British Joint Declaration on Security and Defence, which called for a joint force headquarters and cooperation on the manufacture of unmanned drones, and which sought to further enhance aircraft carrier interoperability.

Reacting to Major Shifts in the International System

Contemporary French national security interests have had to adjust to two main jolts to the international system: the end of the Cold War, and the War on Terror. The former has proved to be a key causal factor behind a significant shift in French strategic thinking that has seen conventional, as opposed to nuclear, military capacity becoming the primary motivation. Indeed, the transformed realities of the new post–Cold War security environment and the potential marginalization of France's role within it were rammed home in dramatic fashion by the country's engagement in the 1991 war against Iraq following the invasion of Kuwait. France's experiences through the first half of the 20th century had spawned an understandable obsession among its national leaders during the second half with territorial defense. This was encapsulated by the predominance of, and upmost faith in, the *force de frappe* program and had as a consequence decades of neglect and underfunding for the country's conventional forces. So, although France was able to contribute quite effectively to Operation *Desert Storm* in the liberation of Kuwait, it was only able to do so in an embarrassingly limited way. It was starkly apparent that a nuclear weapons capability was now of only reduced utility for the attainment of the *grandes lignes* and that emphasis must instead be placed on the capacity to project conventional force abroad. To this end, President Chirac would bring an end to some two hundred years of conscription, or national service, and establish fully professional armed forces.

American actions in the years following 9/11 and the launch of the War on Terror largely flew in the face of French ambitions for the *grandes lignes*. George W. Bush's administration seemingly pushed the United States away from multilateralism and was able to set the agenda (the War on Terror) that the rest of the West had to follow. Two key instruments of France's ability to "control" American policy, NATO and the UNSC, were both marginalized in turn by the White House and the Pentagon. NATO's unprecedented invocation of its mutual defense guarantee in the wake of 9/11 went largely ignored by the United States. Instead, 2002's launch of Operation *Enduring Freedom* in Afghanistan was undertaken with a few select partners (primarily Australia, Canada, and the UK). That said, France would eventually have 5,500 troops deployed in the region. For Tom Lansford, the initial French military commitment might well have been greater had President Chirac's insistence on a multilateral command structure (resolutely blocked by the Pentagon, that was scarred by NATO's 1999 Kosovo campaign, during which strategy and targets had to be approved by all nineteen Alliance members) not been rejected by the Bush Administration.[30] Following victory at the 2012 presidential elections, new French President François Hollande followed up on a campaign promise by announcing that the country's remaining 3,300 troops in Afghanistan as part of ISAF would be withdrawn by the end of the year, a year earlier than the deadline set by his predecessor Nicolas Sarkozy, and two years before the main NATO deadline.

The role of the UNSC as a key instrument used in the pursuit and protection of France's national interests with its pretensions to continued Great Power status has been established previously, and it was briefly noted that the 2002–2003 international row concerning Iraq consequently was a cause for deep concern in Paris. In this instance, from a French perspective, the UNSC was sidelined by the Bush administration, again backed by a few select partners, that was determined to get its way. As a result, UN-based multilateralism, a key aspect of French foreign policy, was bypassed by the U.S.-led invasion. Chirac's resistance to Bush's request for more European troops in Afghanistan, seeing this as an attempt to allow American forces to redeploy to Iraq, would then further dampen Franco-U.S. relations.[31]

Reflections

True to form, the 2012 presidential election campaign saw no contestation of the Gaullist narrative resting on French exceptionalism and the *grandes lignes*, so these are set to remain for the foreseeable future as the fundamental tenets of France's approach toward national security even in the face of powerful financial austerity pressures. In terms of two of the key instruments used to attain these, NATO and the UNSC, and after the low point of 2002–2003, perhaps there are signs of optimism from a French perspective. Recent years have, for example, seen NATO assume the primary role in the military operation in Afghanistan. Meanwhile, in 2011, President Sarkozy, with strong support from the UK and Barack Obama's administration, was able to place the UNSC center stage as it passed a relatively expansive and robust mandate on Libya that was subsequently implemented by NATO and some partners. For a while at least, it is now once again paralyzed over the Syria crisis. The UN and the UNSC were reestablished as the primary arbiter on international peace and stability. And the NATO mission over Libya was the first opportunity for the new fully reintegrated France to take a strong leadership role inside NATO and to be framing the agenda.

Sarkozy only lasted one presidential term, and the Socialist Party is now back in the Elysée Palace for the first time since President Mitterrand. However, the early indications of Hollande's administration point to continuity along the traditional Gaullist principles for national security. Yes, there has been an acceleration of the troop withdrawal from Afghanistan, as discussed previously, but that just reflects a slight tactical shift. The economic pressures on the defense budget will remain for years to come, and there's certainly no big appetite among Hollande's administration to renounce claims to France's Great Power status and all that that entails.

Notes

1. Adrian Treacher, *French Interventionism: Europe's Last Global Player?* (Aldershot, UK: Ashgate, 2003), p. 1.

2. Edward L. Morse, *Foreign Policy and Interdependence in Gaullist France* (Princeton, N.J.: Princeton University Press, 1973), p. 132.

3. Treacher, p. 11.

4. Ibid., 13.

5. James F. McMillan, *Twentieth Century France: Politics and Society 1898-1991* (London: Edward Arnold, 1992), p. 65.

6. Treacher, pp. 15–19.

7. Adrian Treacher, "A Case of Reinvention: France and Military Intervention in the 1990s," *International Peacekeeping*, vol. 7, no. 2, 2000, p. 24.

8. Philip M. Williams and Martin Harrison, *De Gaulle's Republic* (Longmans, London, 1961), p. 21, p. 40

9. Ibid., pp.175–176.

10. Adrian Treacher, "France and Transatlantic Relations," in *The Future of Transatlantic Relations: Perceptions, Policy and Practice*, Andrew M. Dorman and Joyce P. Kaufman, eds. (Stanford, Calif.: Stanford University Press, 2011), p. 96.

11. Pierre Lellouche, "Guidelines for a European Defence Concept," in *Europe in the Western Alliance*, J. Alford and K. Hunt, eds. (London: MacMillan, 1998).

12. John Gaffney, "Highly Emotional States: French-US Relations and the Iraq War," *European Security*, vol. 13, no. 3 (Part One), 2004, p. 267.

13. Joachim Krause, "Multilateralism: Behind European Views," *The Washington Quarterly*, vol. 27, no. 2, 2004, p. 50.

14. Irwin M. Wall, "The French-American war over Iraq," *Brown Journal of World Affairs*, vol. x, issue 2, Winter/Spring 2004, p. 125.

15. Adrian Treacher, '"France and Transatlantic Relations," in *The Future of Transatlantic Relations: Perceptions, Policy and Practice*, Andrew M. Dorman and Joyce P. Kaufman, eds. (Stanford, Calif.: Stanford University Press, 2011), p. 101; David S. Yost, "France's New Nuclear Doctrine," *International Affairs*, vol. 82, no. 4, July 2006, pp. 701–721.

16. *Livre Blanc sur la Défense 1994*, Union Générale d'Editions, Paris 1994.

17. *The French White Paper on Defence and National Security*, www.ambafrance-ca.org.

18. Camille Grand, "The French Case: Livre Blanc sur la Défense et la Sécurité Nationale," *EU-US Security Strategies: Comparative Scenarios and Recommendations*, IAI/UI/FRS/CSIS, May 2011, p. 5.

19. Anand Menon with D. Dimitrakopoulos, and A. Passas, "France and the EU under Sarkozy: between European Ambitions and National Objectives?," *Modern & Contemporary France* 18 (4) 2009.

20. Ben Jones, "Franco-British Cooperation: A New Engine for European Defence?," *European Union Institute for Security Studies Occasional Paper 88*, February 2011, p. 12.

21. Grand, p. 6.

22. Matthew Campbell, "Pedalo Captain Steers France Towards Rocks," *Sunday Times*, 18 November 2012, p. 33.

23. *The French White Paper on Defence and National Security*, www.ambafrance-ca.org.

24. Jonathan Eyal, "France Freezes as Europe Melts," *The Independent*, 3 December 1994.

25. "Defense'" in this context does not imply territorial protection as such but rather military integration, although on intergovernmental lines.

26. Anand Menon, "Empowering Paradise? ESDP at Ten," International Affairs, 85, 2, March 2009.

27. Jones, p. 37.

28. Grand, p. 10.

29. Hervé Morin, quoted in "France Wants Europe to Boost Defence Cooperation," *www.euractiv.com*, October 4, 2010.

30. Tom Lansford, "Whither Lafayette? French Military Policy and the American Campaign in Afghanistan," *European Security*, vol. 11, no. 3, 2002, pp. 136-137.

31. Jolyon Howorth, "Sarkozy and the 'American Mirage' or Why Gaullist Continuity Will Overshadow Transcendence," *European Political Science*, vol. 9, no. 2, 2010, p. 205.

4 German National Security Policy in the Post–Cold War World: An Evolving International Role or a Reluctant Power?

Gale A. Mattox[1]

Introduction

The German approach to its national security can only be described as unique. Few countries have experienced the domestic debate Germany has had over its need to achieve "normality." Given the Holocaust perpetrated by Hitler's Third Reich during World War 2 and the occupation of the country after the war, Germans have debated the future direction of their national security and the appropriate role of the country in the international arena and in Europe almost incessantly since 1945. Their preoccupation with working exclusively within a multilateral framework has set German national security policy apart from other nations. The answer in recent years, particularly since the fall of the Berlin Wall and the German Constitutional Court ruling of 1994 has been, "yes, but . . . " to a larger international role, arguing for a UN Security Council (UNSC) seat in the 1990s while resisting or restricting international involvements more recently. Germany continues to struggle with the question of its identity in the world and, by extension, its national security role. The contradictions inherent in the German role have emerged in its decisions to participate in Afghanistan with the third largest NATO force presence, albeit within unique rules of engagement, while withholding support for the U.S. conflict in Iraq and most recently its abstention on the UNSC Resolution 1973 vote on Libya. This latter decision not to participate in a NATO operation backed by Germany's most important European allies (as well as other non-NATO countries such as Jordan, UAE, and Qatar, and Sweden with the support of the Arab League) underscored the continued German ambiguity over its international role.

The Basis of German National Security Policy

German national security policy is rooted in a series of postwar decisions by the victors and by Germany itself. In essence, the United States, the UK, and France determined the future of the Federal Republic of Germany (FRG), and the Soviet Union closed off the borders of the German Democratic Republic (GDR) and dictated its incorporation into the communist bloc. Few states are as influenced by their war experience as Germany even today; the defeat of Hitler's forces and the Holocaust have permeated every aspect of German life since 1945,[2] creating a German populace very wary of conflict, particularly German military participation in conflict.

For West Germany, the 1949 Basic Law (*Grundgesetz*/Constitution) essentially formed the framework for its foreign and defense policy. Enacted into law in May 1949, the Preamble calls on Germans to "achieve, by free self-determination, the unity and freedom of Germany."[3] With West/East German unification on October 3, 1990 through Article 23, the Basic Law remains the law of the land.[4] Article 1 underlining the priority of human rights and the commitment to international institutions and international law is critical to Germany's national security in both its constitutional and political commitment to multilateral institutions. In this respect, Article 24 states that "(1) the Federation may, by legislation, transfer sovereign powers to international institutions. (2) For the maintenance of peace, the Federation may join a system of mutual collective security; in doing so it will consent to those limitations of its sovereign powers which will bring about a secure peaceful and lasting order, in Europe and among the nations of the world."[5] In this latter respect, Article 25 refers to the priority of international law which "takes precedence over the (federal) laws and directly creates rights and duties for the inhabitants of the federal territory."[6]

In 1955, Chancellor Konrad Adenauer grounded his decision to seek German membership in NATO with this constitutional commitment to international institutions, and it has remained a fundamental tenet of German international relations. Its allies also imposed substantial restrictions when Germany joined international organizations. In joining NATO, Germany renounced the use or development of ABC weapons (atomic, biological, and chemical) and took on a military role focusing primarily on a territorial defense. With the end of the Cold War and the specter of the Soviet threat receding, Germany's national security policy refocused on unification via the transatlantic alliance NATO, the EU, and the UN.

In the post–Cold War, with enlarged NATO membership, a new security paradigm emerged with changes by the European Union. In 1992 with the Maastricht Treaty and later with the Lisbon Treaty, Germany agreed to an EU Common Foreign and Security Policy (CFSP) which subsequently added an explicitly security and defense component (CSDP) that committed it to European common security, a concept articulated in St Malo 1998.[7] The EU further expanded its security competency with a rapid reaction capability, Battle Groups and a European Defense Agency, so that today, two organizations—NATO and EU—form the framework for German national security policy. But with an integrated command, NATO has remained the lead military organization—the "hard power" in the Joseph Nye construct—with the EU focus to date primarily on "soft power."[8]

Since 9/11, the primary threat to the Federal Republic has been from terrorism, and official documents identify terrorism as the major threat confronting Germany. This was not a new phenomenon for a country that had experienced the violence of the Bader Meinhof terrorist group in the 1970s, including the assassination of a Supreme Court justice, among others. But the nonstate nature of the terrorism posed a new magnitude of threat, especially as much of the 9/11 attack was planned from Hamburg. As a result, in November 2001, the German government made the decision to employ forces and humanitarian aid to assist the U.S. in war on terrorism in Afghanistan.

Other challenges have also informed German national security in the 21st century, including the following:

1. Proliferation of nuclear capabilities, a focus of German diplomacy as a part of the UN negotiations with Iran
2. Continued regional stability on the European continent—the Balkans as well as a number of "frozen conflicts"
3. Uninterrupted access to energy resources
4. Maintenance of security in the face of environmental and climate change
5. Establishment of economic stability as the Euro has come under challenge

These mirror broadly those challenges laid out in the 2003 (updated 2008) EU National Security Strategy, *A Secure Europe in a Better World*.[9] While the Germans continue to rely on their membership in NATO, the EU and, to a lesser extent but in a higher regard than many other states, the Organization for Security and Cooperation in Europe (OSCE), policy-makers have gradually begun to carve out a security identity.

The German approach to these challenges is based on a sense of absence of an existential threat as posed by the Cold War Soviet Union. Today Russia is a major energy source, and relations are informed by energy and a relatively high volume of trade. Threats to security are dealt with in a more differentiated way, for instance, negotiations with Iran have sought solutions foremost through diplomacy. While President Bush was discussing a preemptive strike to ensure nuclear nonproliferation, the Europeans advocated a "carrot rather than stick" approach through sanctions. With respect to terrorism, the focus of efforts against potential terrorism has been largely a police matter, a common European approach but substantially different from the U.S. approach. In Afghanistan, the German contribution and support for the U.S. efforts have been significant, but broadly the public concern has been on development assistance, and the German force involvement outside and even within the Northern Command has been constrained, at least until the 2009 Kunduz incident, a turning point in the public perception of the deployment. In tangible terms, this absence of a sense of threat has translated generally into smaller defense budgets falling short of a NATO commitment to 2 percent of GDP. There is simply not a sense of military threat as existed in the Cold War, and public support reflects this shift.

National Security Process

While the German federal president (*Bundesprasident*) is not the figurehead that exists in constitutional monarchies such as the Netherlands and UK, as a parliamentary democracy the president also does not have the powers of the directly elected U.S. president. Importantly, he is charged with representing all Germans once elected and is expected to rise above politics. His speeches are often considered guideposts for the future of the country (i.e., in May 8, 1985, President Richard von Weizsacker reiterated Germany's responsibility for World War 2 and its victims, acting as a moral authority in the midst of a revisionist debate by German historians).[10]

It is the federal chancellor (*Bundeskanzler*)—elected by the parliament—who heads the cabinet and takes responsibility for general national security issues (Article 65, Basic Law). (S)He sets the foreign and security priorities for the government. While in peacetime the minister of defense is commander-in-chief of the armed forces (Article 65a), the chancellor takes command when the country is under attack. In this case, the *Bundestag*—with the *Bundesrat*—declares the country in need of defense (Article 115b). While the states have substantial power on many domestic issues in this federal system, the government has "exclusive legislation on foreign affairs" (Article 73, Basic Law).[11] Within the Parliament (*Bundestag, lower house,* and, only on rare occasions, the *Bundesrat,* upper house) are committees with oversight and legislative powers, but the greater impact emanates from the political parties and through their party structures for foreign and defense policy. Seldom will the current coalition parties vote or publicly diverge from the policies of the seated chancellor, albeit there is an exception for "votes of conscious." In the November 2001 vote offering support for the United States in Afghanistan, Chancellor Schroeder risked his chancellorship with a vote of confidence that succeeded despite eight Green Party coalition members voting against involvement.

Several months after Germany became a NATO member in 1955, the government established a Federal Security Council (*Bundessicherheitsrat* BSR) as it was standing up its first forces since the end of the war; in 1969 it became a committee of the German Cabinet. As in the U.S. case, its creation was by legislation, but it is administered by a civil servant (albeit with direct access to the chancellor) with few staff, in contrast to the U.S. National Security Advisor. Its permanent members include the chancellor, vice-chancellor, ministers of defense (standing chair), foreign affairs, interior, economics, justice, and technology, plus (as in the United States) others as appropriate. It meets informally and can consider broad security issues (i.e., it met following 9/11 to discuss measures for counterterrorism). The Council issues recommendations to the full Federal Cabinet for consideration.[12]

Overwhelmingly, however, the Federal Security Council has primarily become a forum to decide issues dealing with weapons sales. These decisions are taken in secret, but as in a recent case often become—at least in part—public. In mid-July 2011 an intense debate occurred in public and within the Parliament over a (leaked) sale of 200 Leopard tanks to Saudi Arabia. While the government pointed to the sales of munitions and other military hardware by the SPD/Green opposition during its 1998–2005 coalition, SPD parliamentary party head Frank-Walter Steinmeier defended the sales by insisting they had not been to an area of conflict or an authoritarian state (*an Diktatoren*) that represses its population—the required standard to block weapons sales.[13] In a heated parliamentary debate, the opposition pointed to the suspected use of Saudi weapons and armored vehicles to repress the Bahraini demonstrators just weeks before the decision. [14]

In national security affairs, the Supreme Constitutional Court (*Bundesverfassungsgericht*) in Karlsruhe has typically left foreign policy to the politicians, but with a few important exceptions. In the 1970s the court found in favor of the sitting SPD/FDP government on the opening to the East (*Ostpolitik*), whereby tensions were minimized and East-West treaties ensured that any changes in the postwar borders would only occur peacefully. The Constitutional Court also played a critical role in the direction national security policy has taken since 1994, in its ruling on use of German forces outside German territory. Under the ruling, German volunteer forces may be deployed outside German borders under certain conditions—only with the approval of the German Parliament, which determines the length of time, conditions of deployment, and any other requirements. In fact, as a result of this strong role in decisions on troop stationing, the German armed forces are often referred to as a *Parliamentary Army*, and Parliament defines each mission in detail, for instance, meeting annually to approve the mandate for the Afghanistan mission. While this ruling does not extend to forces sent into nonthreatening situations (i.e., for assistance, in a post-earthquake and so forth contingency), it does mean that forces sent into areas of conflict require parliamentary approval.[15]

Several ministries play a role in setting the contours of national security policy. In this respect the ministries of defense and foreign affairs are most prominent and work together with the chancellor. Because the government has consisted of a two/three-party coalition (Christian Democratic Union [CDU] majorities include its CSU Bavarian counterpart) in all but one electoral period since 1949, an unusual aspect is the distribution of major ministries among the parties of the national coalition roughly according to their electoral support. This has translated into the Foreign Ministry normally being held by the smaller coalition party with a correspondingly larger impact on national security policy than would be expected (i.e., during the Cold War the Free Democratic Party [FDP] held the Foreign Ministry for years at a time first as a member of the Social Democratic Party [SPD] coalition and then as CDU/CSU/FDP coalition *despite not winning even one* mandate/seat). The tensions in such a process are obvious but can and have at times provoked discussion and consensus on national security policy; for instance, most recently the coalition agreement in 2010 between the CDU/CSU and FDP contained a sentence proposed by the FDP calling for the withdrawal of tactical nuclear weapons (TNW) from Germany and Europe.[16] This sentence has dictated German policy on the issue of TNW since the 2010 election.

The MoD reflects the German historical experience. Because of the activities of the General Staff under Hitler, there has not been a General Staff; the inspector general is the ranking military authority under a politically appointed civilian minister of defense. Most recently, Minister Karl-Theodor zu Guttenberg initiated reform of the armed forces (*Bundeswehr*) that appeared to set the stage for greater international involvement. The reform is in the process of being carried out by his replacement Minister De Maiziere with the potential for substantial transformation in both personnel and operations. In 2011 the last draftee was inducted into the military and the first voluntary career soldiers were sworn into the force. This process of change from conscription to voluntary service is one of the last among the European states and has a broad impact on German society and armed forces. The draft *Bundeswehr* had not only brought in essence all males into the military, resulting often in many becoming career professionals, but those who chose the alternative option of social service have served at a hospital, orphanage, or other social service institution. As a result, the end of the draft has impacted not only the military but the social service institutions as well. The move to a leaner, expeditionary structured force also reduces the size of the force from 220,000 to 175,000–185,000 troops compared with approximately 495,000 plus reserves in the old West Germany and 170,000 plus in East Germany during the Cold War. It portends a major change for Germany, albeit the direction that change will take internationally remains uncertain, particularly with respect to Germany's future international role. Will Germany's new force prepare it for more involvement in NATO and EU operations or will the planned substantial cuts in force constrain it to territorial defense?

Other ministries play a role, particularly in the broader national security policy process such as the *Bundesnachrichtendienst*, an equivalent to the US CIA. But in a more striking contrast to many other countries, the Ministry for Economic Cooperation and Development is viewed as an integral part of national security policy and assumes a large role generally in decision-making. For instance, the reconstruction mission in Afghanistan held utmost priority for the public as a more critical facet of overall security than is the case in many countries. Consequently, Germany allocates one of the Western industrial countries' highest percentages of GDP (albeit Scandinavian countries rank higher) to developing countries (somewhat below 0.5 percent versus the U.S. under 0.3, both short of the 0.7 percent Millennium Goals for 2015).

National Security Strategy

Unlike many of the other countries in this volume, Germany does not issue a National Security Strategy per se. Rather, periodically the MoD publishes a *Defence White Paper*, most recently in 2006, and in 2011 *Defense Guidelines* [17] that are heavily defense-related in contrast to the government issuing a policy directive more broadly outlining national security. In addition, the Foreign Ministry periodically issues a review of its activities, once a year or more often on the Afghan mission as well as other issues. The lack of a national security policy publication in the form such as it exists in many other countries reflects the complexities in which Germany reviews its security policy and identity in the world as well as the strong role of the Parliament.

The CDU/CSU attempted to jump-start a discussion in Germany with a national security paper in 2008. In their discussion of the effort, Guerot and Korski (European Council on Foreign Relations) identified this attempt as a "sign of Germany's coming of age . . . the culmination of a longer-term development by which Germany's policy on international military operations has changed."[18] They question the absence of a serious discussion of both when Germany will use hard power in international contingencies as well as a better discussion of its role in Europe.[19] The German government has not to date issued a formal national security strategy paper, but discussion has continued and Members of Parliament Philipp Missfelder (CDU) and Elke Hoff (FDP) are on record in favor of such a strategy.

National Security Policy Evolves

Cold War Legacy: The Realist Paradigm

Possibly no other country—except the superpowers themselves—so clearly fits the realist paradigm during the Cold War. Germany was figuratively the battlefield of the U.S.–Soviet Union standoff. In the 1950s the Korean War brought concerns over the adequacy of conventional weapons to stop any Soviet attack; during the Cuban Missile Crisis, there was concern over the potential for a superpower clash in Europe, specifically at the pressure point Berlin. As the superpower contest of wills and standoff in nuclear weapons defined the Cold War, the Germans exemplified the concept of bandwagoning that Stephen Walt has so aptly described by attaching itself to the United States for protection since 1945 and directly in 1955 with its NATO membership.[20] For the Western allies—and even the Soviet Union—this posture suited the need to ensure that German power would not rise again as it had twice in the century. For the Germans, a civilian power paradigm ensured its security while it embedded the country in an alliance. As defined by Hanns Maull, civilian power "promotes multilateralism, institution-building, and supernational integration and tries to constrain the use of force in international relations through national and international norms."[21]

When Saddam Hussein moved his forces into Kuwait in August 1990, a direct abrogation of international law, Germany condemned the actions. But in the midst of the September 4+2 negotiations that restored all remaining sovereignty to Germany and the October 3 unification, Germany was the only NATO member not to contribute military forces to the UN-mandated U.S.-led coalition to free Kuwait. Citing their continued commitment to the civilian paradigm and constitutional restrictions to deploy troops out of area, Germany instead joined Japan and Saudi Arabia in funding the allied efforts. This civilian power paradigm continued into the post–Cold War, but clearly became difficult to maintain as the overriding superpower tensions of the Cold War were supplanted by challenging new threats and, as Maull addressed in 2001, Germany also adapted its forces to the new challenges, particularly terrorism.[22]

Germany focused its attention on unification after 1989 with the fall of the Berlin Wall and subsequent disintegration of the Soviet Union in December 1991. It was a herculean challenge and totally consumed the governmental processes. With unification, the new German eastern states became absorbed into NATO, thereby securing the larger Germany within the Atlantic Alliance. In addition, the eastern states became part of the EU, thus continuing the German commitment to multilateral institutions. Germany began to rethink its broader national security policy—would the larger (population approximately 65 to 82 million) Germany become a more active international player? To that point the armed forces had been restricted by a strict interpretation of the Basic Law (*Grundgesetz)* that deployment of forces outside German territory was not permitted unless to defend against an attack on a NATO member state.[23] For instance, there was limited German participation in 1992 in Bosnia. However, changes were evident when the Germans assisted in the enforcement of the Balkan embargo in the Adriatic with a destroyer and three reconnaissance planes, rather than ground forces.[24]

A Fundamental Shift in National Security Policy: The Constitutional Court Rules

In the period after the Cold War, the opposition at the time (FDP with the CSU) took the issue of the role of the armed forces to the German Constitutional Court. The ruling by the Court issued July 12, 1994 permitted the operation of German forces outside its territory contingent on parliamentary support. (Note that this ruling was further defined in 2005 for German troops in Afghanistan.[25]) The fact that the court approved use of German forces in conflict contingencies is significant, and the 1994 ruling was a turning point in German self-perception of its role as a NATO Alliance member and as an international actor. The ruling more immediately made participation in the conflicts in the Balkans possible, initially in Bosnia-Herzegovina. At the Dayton negotiations in 1995, Germany was an active participant both at the talks and in contributing nearly 4,000 peace-keeping troops to the UN force. In fact, the Parliamentary vote was a strong reaffirmation of the Constitutional Court ruling with 543 in favor (out of a total of 656 members) of participation in the deployment of the Implementation Force (IFOR).

Kosovo proved the larger challenge for the fledging international actor, and public acceptance of the first German use of military force was the subject of intense debate. The German public had supported a UN peacekeeping force in Bosnia, but its inclination to become involved in Kosovo proved more tentative initially and anything but automatic because of the lack of a UNSC resolution. Furthermore, although there was clear evidence of mass and indiscriminate killings that sparked German concern, doubts about German military participation provoked debate, particularly within the Green Party, for the first time by a member of a governing coalition (with a history of an anti-NATO party platform).

Newly elected SPD Chancellor Schroeder and Green Foreign Minister/ Vice Chancellor Joschka Fischer were convinced that German participation was justified despite the public hesitancy. By all accounts Fischer's remarks that of all people the Germans should recognize the need to respond in cases of genocide represented a turning point in the debate. His comments struck a responsive chord for the public. As one journalist commented " . . . (for the Germans) the thought of supporting militarism, on no matter how small a scale, was acutely painful"; but Fischer argued that "Germany had a moral responsibility to help defeat ethnic cleansing in Yugoslavia"[26]—announcing at the Green Party Convention in Bielefeld "Nie Wieder Holocaust." Despite some Green Party member skepticism and rejection by the far left and partly former communist Democratic Socialist Party (PDS), the vote by the *Bundestag* was overwhelming with 500 in favor, 62 against, and 18 abstentions—and German Tornados flew sorties in conflict for the time since 1945.

Out of Europe: German Forces to Afghanistan

With the 9/11 tragedy, German national security policy took another critical shift with deployment to Afghanistan. The response to 9/11 was immediate by the European allies who invoked Article 5 for the first time in NATO's history. Several accounts attribute the swift vote to the leadership of the UK and Germany on the morning of September 12, 2001. As the NATO statement declared: "If it is determined that this attack was directed from abroad against the United States, it shall be regarded as an action covered by Article 5 of the Washington Treaty."[27] Although the United States decided not to use NATO forces and proceeded instead with UK special forces and the Afghani Northern Alliance, within a few months the Bush administration had reached out bilaterally to other NATO members. This decision to bypass NATO initially came in for criticism, but by December, German forces and others were headed for Afghanistan.

Overwhelmingly the issue of German commitment to its NATO allies broadly, and to the United States specifically, left no doubt as to the German decision to participate in the war against the terror that had become the highest priority of the United States and NATO. The NATO multilateral commitment was considered a linchpin to German national security, and there was tremendous support for Chancellor Schroeder's resolution presented to the German Parliament in November after the U.S. attacks. The chancellor put the concept to the test and tied the results to a vote of confidence, forcing the opposition to vote against the resolution but achieving an absolute majority by one more vote than necessary with 336 while also losing a number of

Green coalition member votes. The importance of the vote cannot be underestimated as it sent German forces under fire not only out of NATO territory, as in Bosnia for peacekeeping and even more in Kosovo, but this time out of Europe to South Asia onto territory previously not considered vital for German defense.

The focus of German national policy on the parameters of its defense clearly shifted at this point—as Defense Minister Struck commented—German security would be "defended at the Hindu Kush." Having made this decision, the Germans quickly became the third largest force contributor to the efforts in Afghanistan and at the beginning of 2003 assumed the command of ISAF with the Dutch prior to NATO taking over in August. They have maintained a commitment to Afghanistan even while eschewing the embattled southern region. Further, while maintaining and even increasing the troop levels in Regional Command North, at the same time, the "restrictions/caveats" on the troop movements caused friction with its alliance members, particularly those stationed in the south. However, while some countries had announced troop withdrawals by 2011, the Germans had even increased their numbers to 5,350 by mid 2011 and remained committed to withdrawal of combat troops on the U.S. timeline of 2014, unlike many of the other allies that exited earlier.

The decision to go to Afghanistan has clearly represented elite decision making rather than reflected public opinion. While the German public appears to accept more international responsibilities, it also continues to feel the weight of German history. There is a strong reluctance to be involved in the use of force, but, interestingly, few protest demonstrations have occurred against German forces in Afghanistan. For this reason, Germany has continued to uphold the commitment, albeit on German terms. Perhaps more than any other coalition member, Germany has forged a unique role in Afghanistan with its public viewing its participation in terms of humanitarian relief. German Provincial Reconstruction Teams (PRTs) have been an important priority for German troop presence, rebuilding villages and schools damaged by the Taliban or in the conflict. Only recently, following several tragedies and a number of civilian casualties, has the German public focused on the dangers inherent in sending troops abroad.

"No" to Iraq and Libya

In contrast to Afghanistan, Germany opted not to participate in the Iraq conflict in 2003, a reflection of its "yes, but . . . " approach to national security policy and, perhaps even more, of its struggle to determine the evolving nature of its national security policy. This is not to say that domestic politics did not play a role—often and even usually a determining factor in democratic societies. But in the case of not joining the Iraq coalition, the fact that the German public inclination against involvement was perceived strong enough for the incumbent chancellor candidate Schroeder to decide in the 2002 election to announce publicly his decision not to participate in any coalition against Iraq was in contrast to the long-standing German commitment to multilateralism. Particularly surprising was his announcement that Germany would not participate *even if there was a supportive United Nations resolution* (author's italics), directly in contradiction to its commitment to multilateralism. While the German decision not to go to Iraq surprised many, eventually the failure to find weapons of mass destruction (WMD) relieved even critics of Schroeder's decision that Berlin had taken the right action.

The Germans did not remain alone in the decision. After impressive support for the UNSC Resolution 1441 vote against Iraq's Saddam Hussein in the fall of 2002, the spring debate for the UN coalition to go to war against Iraq united the French, Russians, Belgians, and Germans against war even as the United States led others into conflict without a second authorizing UNSC vote. For the Germans, who coincidently held the UNSC presidency during the February 2003 Iraq debate, the decision not to participate contradicted both its ongoing efforts to secure a permanent seat on the Security Council and its stated commitment to NATO and its allies.

A second seemingly contradictory decision came in March 2011 during the chancellorship of Angela Merkel in coalition with the FDP. As the Arab Spring of 2011 unfolded, the Libyans encountered a highly recalcitrant President Quaddafi who not only refused to step down, as had Egyptian President Mubarak and Tunisian President Ali, but also began to fire on innocent civilian protesters. This led to a request for assistance by Libyan rebels and a vote on UNSC Resolution 1973 to assist the Libyans. In a surprising move, the Germans abstained from support in the vote on Libya. Moreover, they were joined by Russia and China against their NATO allies France, the UK, and the United States. The decision was more than just nonparticipation in Libya;

it required Germany to withdraw its two warships that were stationed in the Mediterranean as part of the NATO war on terrorism, as well as the aircrews Germany contributes to NATO's AWACS force.

These two decisions—one by a Social Democratic Chancellor/Green Foreign Minister and the other from a Christian Democratic Chancellor/Free Democrat Foreign Minister—were clearly a turn away from what had been perceived after the 1994 Constitutional Court decision as a blueprint for a greater international role for Germany. With the decision not to "commit to boots on the ground," in Iraq or Libya, German commitment to a broader international role would have appeared to dissipate. But this would be a questionable conclusion. In both cases, it is true that Germany decided not to participate in a conflict where its major allies were committed. Less known is the participation behind the scenes in both conflicts. In the case of Iraq, German chemical weapons units stayed prepositioned in Kuwait to help with any potential chemical clean-up. In 2011 it assisted with intelligence and other assets that kept them off the ground but contributed support to the allied efforts in Iraq.

In both the case of Iraq in 2003 and Libya in 2011, Germany increased its forces in Afghanistan and continued its support for other ongoing NATO and EU operations, as in Bosnia and off the coast of the Horn of Africa. But these efforts do not negate Germany's "no" to joining those coalitions and underscore the contradictions in the German sense of its national security interests, previously grounded solidly in its alliance commitments and in a multilateral approach. The case of Libya was even more pointed in its abstention and seriously questions the direction Germany intends to pursue with respect to its role in international organizations—many suggested that a "yes" UNSC vote but a decision against contributing forces would have been more appropriate to German commitments.

There remains a strong sense of multilateralism, but do these decisions reflect real shifts in opinion over current security structures? German support for NATO as "essential for their country's security" dropped dramatically from 74 percent in 2002 to 56 percent in 2010.[28] This German figure is surprisingly lower on the same question posed to the general EU public on NATO, which polled 60 and 65 percent when the question was posed to EU leaders.[29] Is the German drop in support rather short term and does it reflect the growing desire to end the war in Afghanistan? In 2010 half the Germans (50 percent) wanted to withdraw troops.[30] The Annual and Interim

Progress Reports on Afghanistan for December 2010 and July 2011 from the Foreign Ministry with the Defense and Economic Cooperation and Development Ministries paint a positive picture of a situation on the ground that will permit the withdrawal of troops, as announced by Foreign Minister Westerwelle, by 2014: "the strategy set forth in the report and the actions of German civilian and military stakeholders have since contributed to the thoroughly positive developments in and prospects for Afghanistan, painful attacks and losses notwithstanding."[31]

This optimism has not convinced the public, but opposition to the conflict has also not to date brought large numbers into the streets as in the 1982 protests against the modernization and deployment of intermediate range theatre nuclear forces.[32] In the case of Afghanistan, the public focus remains on reconstruction and development—albeit ever more cognizant of the implications of military deployments in Southwest Asia. The government recognizes this, and a Bonn Conference reviewing the progress in Afghanistan after ten years (marking a similar 2001 conference also hosted by Germany) laid out the way forward for the region in late 2011. It drew in over seventy-five countries to show commitment to Afghan reconstruction and development in the largest ever government-sponsored meeting held in Germany.

Reflections

In 2010, German President Horst Koehler abruptly stepped down from the presidency after public criticism of his remarks to a reporter with reference to the use of German forces "(it could be that) now and again, in an emergency, military operations could become necessary to safeguard our interests, for example free trade routes, . . . to prevent instability in entire regions that would surely hurt our chances." [33] As reported in the *Atlantic Times*, many Germans criticized the comments as a justification of wars fought for economic reasons. An opposition leader said that Koehler was "endangering acceptance of Germany's foreign missions," and a constitutional law specialist called it "an imperialist turn of phrase."[34] These remarks and the sharp reaction suggest that Germany has clearly not yet come to terms with its role either within the country's elite or in the public. Koehler's comments and unexpected resignation reflect the German struggle today to define its national security challenges and role in the world.

The German broad sense of security through diplomacy, development, and defense has moved the country beyond Hanns Maull's civilian power paradigm of an earlier era. German involvement abroad has grown dramatically, but that involvement until the Kosovo conflict and particularly Afghanistan was by and large within the civilian framework. These have included operations such as rescue of civilians in Lebanon UNIFIL (2006), Sudan UNAMID (2007), AMIS disaster relief/Darfur (2004), Oder (1997) and Elbe Flood Relief operations (2002), Cambodia UNTAC medical assistance (1992/3), earthquake in Morocco (1960), and others.[35] Articulated as far back as the former Chancellor Willy Brandt North-South Commission, German involvement in the international arena has been heavily focused on assistance to the developing world, including even the current Afghan deployment. Only recently has the public—in spite of casualties—focused on the role of the armed forces abroad.

In this context, the discussion of the use of the term "war" has been yet another indication of the complexities of German national security. In September 2009 Colonel Georg Klein observed two fuel trucks in the middle of the night stalled in Kunduz, a town in the Afghanistan Regional Command North spearheaded by the Germans. Believing that the trucks might be set to blow, the colonel called in a NATO air strike, not knowing that many of the town residents had been told that fuel was available and were hurrying to the trucks with pails. German security expert Constanza Stelzenmueller in a *Financial Times* editorial cited a U.S. journalist that it was "the most deadly operation involving German forces since World War 2" and went on to ask "will we also remember it as the night Germany grew up and started to call a war a war?"[36]

This reluctance to call a "war a war" reflects the German culture in the post-1945 era, to attempt to deal with its wartime history and determine its international role in the 21st century. Kunduz is viewed as a turning point in the perception of the Afghan mission, where the German public had to that point focused on the reconstruction of the country and not the security aspect. As a result of the incident, Defense Minister Franz Josef Jung was relieved by Christian Social Member of Parliament Karl-Theodore zu Guttenberg, the former Economics Minister and a rising star in the Conservative ranks. Kunduz became such a debated topic that Mr. Jung stepped down even as the new Minister for Labor and Social Affairs the next month. As Stelzenmueller comments on the issue, the episode "undermine(d) one of Germany's cherished ideas about itself; that, whatever the state of its military, it is at all times morally a superpower."[37] Afghanistan, some had proclaimed, was a "non-international conflict" or only Jung's *Kriegsaehnliche Szenario* (a warlike scenario). The new Defense Minister zu Guttenberg took steps to approach the German role more realistically in April 2010 and pronounced the war "colloquially" a war or a *Kriegsaehnliche Zustaende* (warlike condition). The shift is considered a major step for public perceptions, but the implications remain unclear.

This more realistic conceptualization of German actions in Afghanistan along with the increasing geographical reach of Germany since 1994 has shifted the security paradigm for Berlin, but change is and will remain slow. While Bosnia—where the Germans have been the largest force since the United States downsized to concentrate on Iraq in the mid 2000s—and Kosovo moved the focus of German military involvement for the first time out of the traditional territorial bounds of the transatlantic alliance, Afghanistan marked the first military foray outside of the European continent. But these involvements must be seen against the decision to say no both to participation in Iraq and in Libya and the focus on reconstruction in Afghanistan, particularly until Kunduz in 2009. The *White Paper* issued by the Defense Ministry underscores this uniqueness in the German approach to security in its first paragraph outlining the challenges to the country—"a successful response to these new challenges requires the application of a wide range of foreign, security, and defense and development policy instruments."[38] While the United States is moving to the inclusion of development in its broad concept of security, this "civilian" factor has long been a major component for German policy.

Furthermore, the continuing shortfall in willingness to make expected military contributions to the Atlantic Alliance[39]—as U.S. Secretary of Defense Robert Gates commented in his farewell speech in Brussels—as well as European defense efforts are contradictory about the emerging international role of Germany that Berlin signaled in the late 1990s/early 2000s with its attempts to secure a UNSC permanent seat. In May 2011 the new Defense Guidelines proclaimed "Assuring the National Interests—Taking over International Responsibility—Building Security Together."[40] The contradictions between this and the abstention one month prior on UNSC Resolution 1973 with respect to the "responsibility to protect" Libya is obvious and leaves the question of the future direction of German security policy in the context of the international arena still an uncertainty.

Notes

1. The views expressed in this chapter are those of the author and do not reflect the views of any government affiliation or other institution. The author would like to

thank the U.S. Naval Academy Research Council in summer 2010 and 2011 and the USAF Institute for National Security Studies for its travel support in summer 2011 to undertake the research for this chapter.

2. There continues to be a debate over the impact of the war even today on the Foreign Ministry and its diplomats. See Norbert Frei, "Das Amt und die Vergangenheit: Deutsche Diplomaten im Dritten Reich und in der Bundesrepublik," Report of the Independent Historians' Commission, Leo Beck Institute, New York,/Berlin 2011, p. 17 (monograph).

3. U.S. Department of State, "Basic Law (Constitution) of the Federal Republic of Germany, Approves by the Parliamentary Council in Bonn, May 8, 1949," *Documents on Germany 1944–1985*. Department of State Publication 9446, Office of the Historian, Washington, D.C., p. 221. In a 1994 Forward to the Basic Law, Federal President Herzog wrote an introduction including "*On 3 October 1990 Germany achieved national unity. By virtue of a sovereign, conscious decision of the people, the Basic Law became the constitution for the whole nation. . . .*" http://www.iuscomp.org/gla/statutes/GG.htm#Preamble.

4. This commitment to unification pervaded West German foreign and defense policy, i.e., in 1968/69 the Nuclear Nonproliferation Treaty was only ratified by the German Parliament with the agreement that the treaty would include eastern Germany should unification occur at a future date. The Basic Law after 1990 included the following from Federal President Herzog: ". . . *The fact that only adjustments were necessary attests to the Basic Law's excellent quality as the foundation of our polity from its inception. . . .*" (1994). http://www.iuscomp.org/gla/statutes/GG.htm#Preamble.

5. U.S. Department of State (Basic Law), p. 226.

6. Ibid., 227.

7. Franco-British Summit Joint Declaration on European Defense. St. Malo, 4 December 1998. http://www.atlanticcommunity.org/Saint-Malo%20Declaration%20Text.html.

8. Joseph S. Nye, *The Paradox of American Power: Why the World's Only Superpower Can't Go It Alone* (Oxford University Press, 2003). See also Nye's subsequent books as well as discussion on hard, soft, and, most recently, smart power.

9. EU National Security Strategy Brussels, 2003/updared2008. *A Secure Europe in a Better World—The European Security Strategy* http://www.consilium.europa.eu/uedocs/cmsUpload/78367.pdf. Note that the EU also included organized crime, which has been expanded upon in an EU Internal Security Strategy released spring 2010.

10. Richard von Weizsaecker, German President. *Speech in the Bundestag on 8 May 1985 during the Ceremony Commemorating the 40th Anniversary of the End of War in Europe and of National-Socialist Tyranny.* English translation. http://www.mediaculture-online.de.

11. U.S. Department of State (Basic Law), p. 237.

12. Authored by OTL iG Christian Behm, Department of WD 2, foreign affairs, international law, Economic Cooperation and Development, Defence, Human Rights and Humanitarian Aid based on web site (below) drawn from a number of sources: Michael Hollmann and Kai von Jena (1955). The Cabinet minutes of the Federal Government, vol. 8, pp. 552–568; Jan Kuebart (1999), Federal Security and Federal Security Bureau, in *European Security* 48, pp. 40–43; Kai Count, (2005), The Federal Security Council, in *The State* 44, pp. 462–482; and various publications/speeches from the web site of the federal government under http://www.bundesregierung.de / Web / Breg / DE / service / search / Functions / filter form, templateId = processForm.html. 13. Cordula Eubel and Stephan Haselberger, "Interview with Frank-Walter Steinmeier: Es gibt eine Sehnsucht nach Fuehrung," *Tagesspiegel*, July 24, 2010, p. 7.

14. Ibid. When asked specifically about the sale of patrol boats during the SPD government, Steinmeier responded that the sale did not undermine the standards—Angola was not an area of conflict nor did the patrol boats have the potential to be used against the Angolan populace, only to protect the livelihoods of Angolan fishermen and the Angolan food supply.

15. Dieter Dettke. *Germany Says "No": The Iraq War and the Future of German Foreign and Security Policy* (Baltimore: Woodrow Wilson Center for Scholars Press/Johns Hopkins University Press, 2009). See also for discussion of the Constitutional Court case in which the SPD/FDP asked for clarification on two resolutions dealing with the use of German AWACS over Bosnia-Herzegovina and a Bundeswehr transport battalion in Somalia, both in 1992. For discussion of the 12 July 1994 decision, see Georg Nolte, "Bundeswehreinsaetze in kollektiven /Sicherheitssystemen—Zum Urteil des Bundesverfassungsgerichts vom 12. Juli 1994," 54 *Zeitschrift fuer Auslaendisches Oeffentliches Recht und Voelkerrecht*, 652, 673 (1994). BVerfGE 90, 286, 383–384.

16. Estimated at two hundred in Europe, these nuclear weapons are B61-type freefall gravity bombs stored in bunkers that would be mounted on and dropped by dual capable F-16 or Tornado aircraft.

17. German Federal Ministry of Defence. *White Paper 2006 on German Security Policy and the Future of the Bundeswehr*, 2006 and German Ministry of Defense, "Defense Guidelines," Berlin, 27 May 2011, 20 pp. See German Ministry of Defense web site.

18. Ulrike Guerot and Daniel Korski, "New German Security Strategy—Going It Alone?" European Council on Foreign Relations, May 15, 2008. Paper version.

19. Ibid., Guerot/Korski.

20. Stephen Walt, *The Origins of Alliances* (Ithaca, N.Y.: Cornell University Press, 1987).

21. Hanns W. Maull, "Germany and the Use of Force: Still a Civilian Power?," Trierer Arbeitspapier zur Internationale Politik (Trier Working Paper on International Politics), Nr. 2, November 1999, p. 1. Prepared for a Brookings Institution Workshop July 1999.

22. See also Hanns Maull, "Zivilmacht Bundesrepublik Deutschland—Vierzehn Thesen fuer eine neue deutsche Aussenpolitik," *Europa Archiv*, 43, 1992, pp. 269–278. Also Sebastian Harnisch and Hanns W. Maull, *Germany as a Civilian Power? The Foreign Policy of the Berlin Republic* (Manchester, UK: Manchester University Press, 2001).

23. See Article 24, para 2 of the *Basic Law*. "with a view to maintaining peace, the Federation may enter into a system of mutual collective security; in doing so it shall consent to such limitations upon its sovereign powers as will bring about and secure a lasting peace in Europe and among the nations of the world." Furthermore, Article 87a states that "apart from defense, the armed forces may be used only to the extent explicitly permitted by this constitution."

24. For an excellent history of German operations, see Rainer Baumann, "German Security Policy within NATO" in *German Foreign Policy since Unification*, ed. Volker Rittberger, Issues in German Politics (Manchester, UK: Manchester University Press, 2001).

25. The Parliamentary Participation Act of March 24, 2005, outlined the procedure by which the *Bundestag* (parliament) could mandate a mission. The mandate is for twelve months, although a mission may be withdrawn early. It may be renewed without a renewal vote only if the "content of the mission has not changed." The motion on a mission must include the following: "operational mission, area of operation, duration, costs, legal bases, maximum number of soldiers required, as well as their skills. Parliament may then approve or reject the German Government's motion by a simple majority. It cannot make any changes to the motion." Regular reports must be made on the mission during the twelve-month periods. Http://www.bundesregierung.de/content /EN/Artikel/Auslandseinsaetze-der-bundeswehr/ (legal bases for *Bundeswehr* missions abroad). Note that missions to extricate people from dangerous situations may be reported after the fact, and humanitarian missions do not require a parliamentary vote.

26. David Glenn. "The Greens' War: A Divided German Left Opts Into the Afghan War," *The American Prospect*, December 10, 2001: 1. Tornados had flown reconnaissance flights over Bosnia in 1995 according to a report by NDR Radion bu journalist Klaus J. Halle.

27. Edgar Buckley. NATO Assistant Secretary General for Defense Planning and Operations 1999–2003. "Invoking Article 5," *NATO Review*, Summer 2006. Web: http://www.nato.int/docu/review/2006/issue2/english/art2.html. Buckley maintains that the statement written by him and his staff was adopted directly by the North Atlantic Council on September 12, 2001.

28. German Marshall Fund, *Transatlantic Trends: Key Findings 2010*, Washington, D.C., p. 18.

29. German Marshall Fund, *Transatlantic Trends: Leaders. Key Findings 2011*, Washington, D.C., p. 11. Note that the U.S. general public paralleled that of the EU public with 60 percent agreeing that "NATO is still essential." However, U.S. leaders led EU leaders 76 to 65 percent on the same question.

30. German Marshall Fund, *Transatlantic Trends: Key Findings 2010*, Washington, D.C., p. 18.

31. Ambassador Michael Steiner, Federal Government Special Representative for Afghanistan and Pakistan, Foreign Ministry, *Progress Report on Afghanistan for the Information of the German Bundestag*, Interim Report, Berlin, July 2011, p. 1.

32. Gale A. Mattox, "The United States Tests Détente" and Samuel F. Wells, "From Euromissiles to Maastricht: The Policies of Reagan-Bush and Mitterrand," H. Haftendorn et al., ed., *The Strategic Triangle: France, Germany, and the United States in the Shaping of the New Europe*, Johns Hopkins University Press and Woodrow Wilson Center Press, 2006: 261–308.

33. Lutz Lichtenberger, "Highest Office in the Land," *The Atlantic Times*, July/August 2010, p. 5.

34. Lichtenberger, p. 5.

35. See additional deployments in Gale A. Mattox, "Germany: From Civilian Power to International Actor," in Andrew Dorman and Joyce Kaufman, eds., *The Future of Transatlantic Relations: Perceptions, Policy and Practice*, Stanford University Security Series, 2010, p. 130.

36. Constanze Stelzenmueller, "Germany Shoots First and Thinks Again," *Financial Times*, September 9, 2009, p. 8. Approximately 142 Afghan civilians were reported to have been killed by the USAF F-15Es that were called in by the commanding officer Colonel Klein, not realizing the presence of the civilians. Defense Minister Jung at the time minimized the casualties and came under severe criticism.

37. Stelzenmueller, p. 8.

38. Federal Ministry of Defense. White Paper 2006. p. 20.

39. U.S. Defense Secretary Robert Gates to NATO, Brussels, 10 June 2011. Transcript available on U.S. Department of Defense web site. "In the final analysis, there is no substitute for nations providing the resources necessary to have the military capability the Alliance needs when faced with a security challenge. Ultimately, nations must be responsible for their fair share of the common defense."

40. German Ministry of Defense, "Defense Guidelines," Berlin, 27 May 2011. 20 pp. See German Ministry of Defense web site.

5 National Security and the United Kingdom

Andrew M. Dorman

THIS CHAPTER EXAMINES THE CASE STUDY OF THE UNITED KINGdom (UK) and its approach to national security. The United Kingdom provides a fascinating example of a state that was once a great empire, comprising over a quarter of the world's population and with the world's largest economy, whose relative power and standing within the international system has been in a state of relative decline for most of the last century. It is now attempting to confront a rapidly changing situation in which its political elite feel constrained by the historical legacy with which they have been left.[1] In this respect, the United Kingdom is similar both to France and Russia. The challenges confronting the British government have been exacerbated by the 2008 financial crisis, which particularly affected the United Kingdom with its dependence on the financial sector and traditional U.S. and European markets.

Yet despite all of this, the UK still remains a permanent member of the UN Security Council; it is one of the leading members of NATO with a full range of military capabilities; is a somewhat awkward member of the European Union; a member of the Five Eyes defense and intelligence community; and the leading member of the Commonwealth of Nations. It retains the ability to project military power, albeit at a reduced level, globally and if necessarily independently as the 2000 deployment to Sierra Leone showed.[2] Moreover, in London it houses one of the three leading world stock exchanges, it is home to a number of the world's largest financial and banking companies, and it is also home to the main shipping insurance market.

The chapter has been subdivided into four parts. The first part addresses the issue of identity, which continues to constrain policymakers. The second part examines the evolution of the idea of the national interest in the UK, including the systemic issues being currently confronted. The third part then analyzes how the successive governments have sought to support the national interest, paying particular attention to the importance of alliances and partnerships, changes to the UK's national security apparatus, and how the UK has chosen to articulate its national security. The final part makes a number of reflections on the UK's national security.

The Problem of Identity

In any examination of the UK, one is immediately confronted by the conundrum of identity. Even its current name—the United Kingdom of Great Britain and Northern Ireland—represents part of the evolution of the state, having been formerly changed in 1927 from the United Kingdom of Great Britain and Ireland as a result of the creation of the Republic of Ireland. It therefore identifies itself either as the United Kingdom or Great Britain and Northern Ireland, and this complex identity has been used to its advantage in international meetings with the choice of identity chosen based on where it is deemed most advantageous to sit. Thus, in forums in which the United States is present, the British representative invariably sits as the representative of the UK and thus next to his/her American counterpart as a means of exerting maximum influence.

Thus, it is hardly surprising that the UK is frequently referred to as Great Britain, Britain (i.e., a combination of England, Scotland, and Wales), or by one of the four states that comprise it—England, Scotland, Wales, and Northern Ireland. Hidden behind this confusion is the question of identity. Formally, the UK comprises the four nations, affiliations, and fourteen dependent territories, but the reality is in fact a far more complicated and evolving picture.

This confusion over identity is perhaps best exemplified in the sporting arena. All four nations have their own football teams recognized within the Federation Internationale de Football (FIFA), and the three Celtic nations have resisted calls for a single UK team. As a result, there had not been a

UK football team at the Olympics until the 2012 games held in London, at which point a single team was entered by the English Football Association, a team containing representatives of the other nations. The four nations also have separate football leagues, although there are a number of Welsh football teams in the English League. Similarly, each nation has a separate team in Rugby Union, although in this case there is a united Ireland team comprising players from both north and south and thus comprising individuals from two different states.

The identity issue is further complicated by the fact that the three Celtic nations—Scotland, Northern Ireland, and Wales—have different formal relationships within the UK and, most importantly, with England, which remains the dominant nation by virtue of its relative size and having the capital (London) in its territory. See Table 5.1 for the 2011 census population breakdown.

All four nations have their own idiosyncrasies that influence attitudes toward national security. For example, Northern Ireland, which is not part of Great Britain, probably focuses attention more on the Republic of Ireland than it does the other nations of the UK. The so-called "troubles" between the Protestant and Catholic communities have dominated its politics, and their legacy still overshadows their communities.[3] At the height of the troubles, some 28,000 British Army personnel were deployed to Northern Ireland in support of the civilian authorities and along with the police and civilians suffered significant casualties in a conflict that neither side was able to win.[4] The peace process that followed has resulted in the majority of the sectarian violence ending and the withdrawal of the army from direct support to the civilian authorities. A much reduced army presence remains in barracks and performs the tasks it does elsewhere in the UK, but concern remains that some of the dissident groups have not relinquished violence, and there still have been the occasional deaths. Moreover, divisions over whether the Union flag—the flag of the UK—should be flown over Belfast City Hall in January 2013 reflected the deep divisions that remain within Northern Ireland and led to repeated nights of rioting, in which hundreds of police officers were injured.[5]

Scotland and Wales are far more integrated than Northern Ireland within the UK, and each has its own forms of government. Unlike Northern Ireland and Wales, Scotland was not occupied by England. Instead, the kingdoms of England and Scotland were united in the 17th century under James I (of England, VI of Scotland). Yet, it is in Scotland where the issue of independence has moved the most with the ruling Scottish Nationalist

TABLE 5.1 Population Breakdown

Nation	*Population*	*% of total*
England	53,012,456	83.9
Northern Ireland	1,820,863	2.9
Scotland	5,295,000	8.4
Wales	3,063,456	4.8

Source: "Table 2 2011 Census: Usual resident population and population density, local authorities in the United Kingdom" (PDF). Office of National Statistics. 2012. Retrieved 19 December 2012.

Party (SNP) pledging to hold a referendum in 2014 regarding Scottish independence. The agreed question—should Scotland be an independent country?—is likely to result in one of three options: full independence via the dissolution of the Act of Union; Maximum Devolution; and retaining the status quo.[6] The likely result of this vote remains unclear, and sensitivities are clear as the backlash to *The Economist's* 14–20 April 2012 edition demonstrated.[7] Currently, the SNP have pledged to retain the same monarchy albeit with the crowns separated, and remain in NATO and the European Union but retain a common currency with the remainder of the UK.[8] Critics have questioned whether these policies would or could work in practice. For example, legal opinion given to the UK government has indicated that the remainder of the UK would inherit the UK's seat on the UN Security Council and that Scotland, as a new nation, would have to renegotiate all its treaty commitments, including applications to join both NATO and the European Union.[9] If membership of the EU had to be applied for then Scotland would have to accept the Euro as its currency and abide by the Schengen Agreement, which would lead to border control between Scotland and the rest of the UK. NATO membership seems more doubtful given the SNP's rejection of the Washington Treaty provisions and the potential loss of the UK's nuclear commitment to NATO, resulting from the loss of Scottish nuclear bases

Wales is the nation most integrated with England, partly as a result of having been occupied by England for the longest period. Nevertheless, it retains its own identity and a separate language. In recognition of the closeness of the two nations, the heir to the English throne (currently Prince Charles) is by convention bestowed with the title Prince of Wales. Thus Wales is in a more junior position in its relationship with England compared with Scotland, and this is most evident in the fact that Scotland has greater devolved powers and its own legal system.

As a consequence of all, there are direct consequences for national security. For example, the composition and deployment of Britain's armed forces are skewed to reflect the four nations. Thus in the 2004–2005 British Army reorganization, plans originally included the merger of the five guards regiments (two English, one Welsh, one Scots, and one Irish), but this was blocked by the queen to ensure that each nation was represented in the armed forces.[10] Similarly, in the army's Future Force 2020 reorganization, which was announced in 2011, the Scottish line regiments were protected. Instead, cuts were made to one Welsh and three English battalions despite the fact that the Scottish battalions were the worst recruited in the British Army and were dependent on recruits from the north of England and the Commonwealth to make up their shortfalls. In a similar move to placate all three Celtic nations, the reduction to seven Infantry brigades will see one headquartered in each of the Celtic nations and only four in England, despite England representing over 80 percent of the population.

Interestingly, the increasing devolution of power to the Celtic nations and the calls for a repeal of the Act of Union between England and Scotland have contributed to the English voice beginning to be reemerge. Once again the flag of St. George, which had become the preserve of the far right, has become far more prevalent and acceptable within the political mainstream. Although there have been no significant calls for a separate English parliament or assembly (the trial of a regional body for the North-East of England proved a dismal failure),[11] there have been calls for the Members of Parliament representing Celtic constituencies to be barred from voting on legislation that applies solely to England—the so-called West Lothian question. This was particularly noticeable under the previous Labour government, which relied on its seats in Wales and Scotland to give it the parliamentary majority to govern. These calls have been less frequent recently, mainly because the present coalition government is dominated by the Conservative Party, which is essentially an English party having lost all of its seats in Scotland bar one. Moreover, coalition plans to reform the House of Commons involving reductions in the overall number of Members of Parliament and making each constituency equal in population size (with the exception of the Highlands and Islands seat, which will have roughly half the normal electorate because of its geographical size) will increase the proportion of English seats in the House of Commons.[12]

Should the Labour Party win the next general election and Scotland stays in the union, then there is likely to be increased calls for baring the Members of Parliament from the Celtic nations from voting on issues relating to England.

The UK also retains responsibility for fourteen overseas territories that reflect the legacy of its imperial past. These are scattered across the globe from the Falkland Islands in the South Atlantic to the Gibraltar in the Mediterranean and Pitcairn Islands in the Pacific. Their sense of identity, and particularly the notion of Britishness, can at times be more vocal than that in the United Kingdom. This is no more evident than in the Falklands and Gibraltar, which have sought to avoid becoming part of Argentina and Spain, respectively. With a recent conflict over the Falklands in 1982, the responsibility for their protection forms a core part of British national security policy.

Of perhaps greater significance is the continuing changing ethnic makeup of the UK.[13] Like America, the UK has been a land of immigration for centuries as various disaffected economic and political disaporas fled to and settled there. This continues today, with one in four babies born in England and Wales having mothers born outside of the UK.[14] At the same time, the UK remains a land of significant emigration, with an estimated 8 million British citizens living abroad. This means that who comprises a citizen of the UK remains in a state of flux. This is perhaps best demonstrated in two examples. In London, there are approximately forty large communities of people born outside the UK, and Londoners speak more than three hundred different languages.[15] In other parts of the UK, individual ethnic groups have frequently concentrated themselves in particular areas, retaining close links with the countries they were historically associated with. Similarly, British citizens living abroad in places such as Spain have elected their own representatives to local and regional governments.

The Evolution of National Interest

This issue of identity is compounded by the UK's history and its focus on the past. In his 1987 book *The Rise and Fall of Great Powers*, Paul Kennedy argued that the UK was another example of an empire that had risen and then ultimately fallen as a result of over-extension.[16] This declinist view has dominated much of the thinking within the academic literature and the understanding of the political elite about the UK's position in the international system. As a result, British external relations from the 1950s have been dominated by the idea of maintaining "Britain's place in the world." Thus the emphasis of successive governments has been on managing the decline and retaining influence rather than looking ahead with a vision of where the UK should fit within the international system.

Interestingly, the term national interest has not generally been included within the language of government publications until fairly recently. In part, this would seem to be because the political elite and the officials seemed to think that national interest was a concept with negative nationalistic connotations and instead preferred to elucidate policy in terms with a positive narrative such as "forces for good." This has been particularly so for Labour governments, which have sought to appease their antinuclear left wing by appealing to their New Internationalists credentials. For Conservative governments, references to the national interest have far greater resonance. Although the two main political parties—Labour and Conservative—have tended to use different language to describe their respective foreign policies, in reality both have tended to follow a fairly similar path. Under Tony Blair's leadership there was a coming together of the respective narratives in which national interest was directly linked to the need for "liberal interventionalism."[17] In opposition the current Conservative government sought to place far greater emphasis on national interest and implied that the decade of liberal interventionalism that had seen the long-term deployment of the British armed forces to Iraq and Afghanistan was over. However, the recent Franco-British led NATO operation in Libya, Prime Minister Cameron's subsequent speech to the UN General Assembly in 2011, and now the British military support for the French deployment to Mali appear to have breathed new life into the ideas of liberal interventionalism.[18] In fact, the government's use of language used to accompany the British support to the French Mali deployment was almost Blairite, and the Labour opposition are now talking about preventive interventions.[19]

Nevertheless, the UK's national interest has been fairly consistent for a number of centuries and comprises two elements. The first element that has remained unstated has been the preservation of the union of the four nations, while a second element has been the protection of the UK's interests as a trading nation. This is a result of the relative density of its population and lack of all the necessary raw materials to sustain it. To survive, the UK has to trade with other states, and the protection of trade has been the dominant external requirement for centuries, which has meant that it has always sought international engagement and cooperation. As a result, to support this second element four interlinked assumptions have emerged and remained consistent. First, there has been a consistent goal of avoiding a single power or group of powers dominating Europe and thus potentially disrupting Britain's trade with its largest market.[20] During the Cold War this focused on the Soviet Union; more recently it has involved ensuring that no European state or group of states dominates either NATO or the European Union.

Second, there remains a continued emphasis on Britain's so-called special relationship with the United States. Although the notion of a special relationship has been much maligned within academia and the press, its origins can be found in the partnership forged during the World War 2 and the shared cultural heritage, most notably a common language. The United States' emergence as a superpower and the UK's relative decline meant that successive British governments saw the United States as the only potential counterweight to the Soviet Union in the immediate aftermath of the World War 2. With the demise of the Soviet Union, this emphasis has continued as the United States assumed the mantle of sole superpower. Furthermore, while Europe has been the UK's largest market, the United States has been Britain's single largest trading partner, and for many years it has been the number one investor in the UK economy and the country of choice for British overseas investment. This is most evident in the financial market, where the 2008 banking crisis explicitly highlighted how intertwined the U.S. and UK financial sectors were.

Third, there has been the assumption that ultimately the only guarantee of a state's survival was by the retention of its own nuclear deterrent. In the immediate aftermath, this was seen primarily as a mechanism of deterring the Soviet Union should the U.S. defense guarantee to the UK and the rest of NATO appear to lose credibility. This assumption stems from the UK's experience of May 1940 and the Battle of France, where the British resisted sending additional fighter squadrons to France and instead retained them for its own continued defense. This led British policy-makers to conclude that ultimately no state will commit suicide for another. Thus, in the early years of the Cold War the United Kingdom developed its own force of nuclear bombers—the V-Force—but from the late 1950s onwards the British nuclear deterrent, as it is known, became increasingly interwoven with the United States to the extent that the UK is dependent on the United States for the provision of its nuclear delivery system—the Trident submarine-launched ballistic missile.[21] Although the threat posed by the Soviet Union has disappeared, and the current assumption is that no state threatens the UK with nuclear weapons, the continued retention of a nuclear capability has remained part of government policy, with the present coalition government committed to replace the existing submarines in the next decade.

Finally, to help protect Britain's interests, successive governments have sought to position the UK to maximize its ability to influence decisions on the world stage through its membership of key organizations such as the UN Security Council, NATO, the EU, and the Commonwealth.[22] This links directly to the issue of identity, with emphasis continuing to be placed on what the UK was rather than any particular vision of what the UK would like to become. However, the UK is now confronted with four main challenges. First, the rise of the so-called BRIC nations (Brazil, Russia, India, and China) and the shift of the world's economy from the transatlantic region toward Asia-Pacific are seen as threats to Britain and Europe's relative standing in the world. This directly links into the continuing declinist agenda and is compounded not just by the relative rise of these states but also by the U.S. reaction to them. There is considerable fear that there is a U.S. pivot toward the Pacific, which might lead the United States to abandon or neglect its transatlantic partnership. This is perceived to be directly contrary to Britain's national interests.

Second, this situation has been exacerbated by the changes in Britain's economy. The last thirty years have witnessed a significant decline in Britain's manufacturing base and a shift toward a service sector economy, with banking and finance being its major outputs.[23] This left the British economy particularly vulnerable to the impact of the financial crisis in 2008. As a result, the then Labour government was forced to intervene and acquire the majority of shares in a number of banks, with the result that the UK's national debt has soared beyond the £1 trillion level.[24] The situation has not resolved itself, and the UK's dependence on trade, particularly with Europe, has meant that it remains deeply concerned about the future of the Euro and any potential impact that might occur from one or more of the European states defaulting on their debt. In particular, the UK was quick to act in support of the Republic of Ireland because of the potential impact on a number of its banks and also on the peace process in Northern Ireland. Linked to this economic point are the countries where the UK exports to. The importance of the Republic of Ireland for British trade is reflected in British exports to the Republic of Ireland being more than those to Brazil, China, and India combined. More widely, the UK's major export market remains

Europe but with the largest single nation being the United States. This, in part, explains the drive of the coalition government to expand its market share in the Asia-Pacific and South American areas. Whether it will be able to compete with the low-cost economies of countries such as China remains to be seen, and the British economy is still struggling to pull out of recession and return to even its 2008 level.

Third, the advantage of the English language as the principal language of commerce has also meant that for economic migrants the UK continues to present itself as a potential source of opportunity. This has led to a constant net influx of people both from the EU and elsewhere. If this trend continues then the growth of the British population in comparison to its main European partners—France and Germany—will alter the relative balance between them and raise significant societal issues for the UK as it tries to manage the needs of its citizens.[25] The UK, and Europe more generally, is confronted with the very real prospect of the UK replacing Germany as the largest European country in terms of population and therefore becoming the leading nation within European (if Russia and Turkey are not defined as European). Thus, the nation that sees itself as being separate from Europe will become Europe's most populous nation, which raises quite profound questions for how the UK engages with the EU. A possible vote on Britain's continued membership of the EU looks likely to happen if the Conservatives are elected to office in 2015. Public opinion has generally been anti–European Union, and thus there is the prospect of the EU's second most populous member (and likely to become most populous) leaving the EU. No mechanisms have been established for this, and the potential impact of such a move has already led the Obama administration to express concern.

Finally, the last challenge for the UK has been to its very existence as the individual members of it grow increasingly apart. In raising the prospect of repealing the act of union, the SNP has raised the question of whether the Union itself will remain. The move toward increasing the relative power of the three Celtic nations has raised the whole identity issue within British politics. This issue of identity has been mirrored within some of the communities that comprise its citizenship and has been most openly shown in the July 7, 2005 attacks on London by four British suicide bombers and the failure of a number of subsequent plots.

Supporting the National Interest

As a result of the UK's perceived national interests, partners and allies have long been a major part of its overseas policies. British foreign policy in the century leading up to the World War 1 saw successive British governments emphasize the maintenance of a balance of power in Europe. With its naval dominance and industrial might the UK saw itself as a preserver of peace in Europe, utilizing a succession of changes in alliance with the UK joining the weak power bloc as a counter to any state or group of states gaining dominance over Europe. This emphasis on a stable, peaceful Europe was purely for practical gain. It then allowed British governments to fully participate in the acquisition of colonies, resulting in Britain becoming the colonial power dominating over a quarter of the world. The reasoning behind the policy is important. While Britain sent missionaries across the globe, its focus was not on the export of its values or beliefs but instead on economic advantage. In fact, the empire was at times used as a safety valve in which to allow those most at odds with the then British system of government and limited democracy to seek alternative places to follow their values (such a policy can be dated back to the Puritans setting sail for America in the 17th century).

However, Britain's industrial and naval dominance began to be challenged in the years immediately preceding the World War 1. First America and then Germany overtook it in terms of economic might while it surrendered its naval ascendancy when it signed the Washington naval agreements of 1922, thus reducing its navy to the size of the United States.[26] Since then, the UK's emphasis on alliances and partners has moved away from that of the world's policeman to one in which it has developed these agreements to protect it from perceived threat or as a mechanism of preserving its interests. During the Cold War this involved not just NATO and the Western European Union but also the Central Treaty Organization that covered the area from Turkey to Pakistan until the 1970s as well as the Five Power Defence Agreement from 1970, which focused on the Pacific (including Australia, New Zealand, Malaysia, and Singapore). The basic assumption for successive British governments has been that military operations as part of a coalition remain preferable, although there have been exceptions such as the Falklands Conflict in 1982 and Sierra Leone in 2000.[27] However, when it engages in operations on a unilateral basis it invariably ensures that it has the support of the United States and others.

As a consequence, the UK continues to maintain a web of alliance commitments and partnerships across the globe. This includes membership of NATO and the EU but also as being part of the Five Eyes intelligence agreement with the United States, Australia, Canada, and New Zealand and the Five Power Pact in the Pacific with Singapore, Malaysia, Australia, and New Zealand. This means that the UK maintains the ability to project limited military capability across the globe with permanent forces stationed in Brunei, Germany, Gibraltar, Cyprus, and the Falklands.

Successive British governments have assumed that the UK is particularly good at a joined-up government approach to policy using the Cabinet Office as a mechanism of ensuring cross-government coordination. However, operations in Iraq and Afghanistan showed that far from adopting an integrated approach, the existing system has frequently seen the various departments of state at odds with one another.[28] Perhaps more fundamental has been a questioning of the British government to think and act strategically.[29]

As a result, the new coalition government has reorganized the UK's security apparatus. A National Security Council (NSC) modeled on the U.S. NSC, which itself had been based on the previous British Committee of Imperial Defence set up in 1902, was formed in May 2010. The new NSC meets at least weekly and is chaired by the prime minister and includes his Liberal Democrat deputy Nick Clegg, the Secretaries of State for Foreign and Commonwealth Affairs, Defence, Home, Energy and Climate Change, and International Development as well as the Chancellor of the Exchequer and his Liberal Democrat deputy. In addition, there are a number of officials, including the new National Security Advisor (NSA), the Chief of the Defence Staff, and the heads of the three intelligence agencies (MI5, MI6, and GCHQ). Unlike the U.S. system, the NSA is an apolitical figure, the first being Sir Peter Ricketts, who had been the principal secretary at the Foreign and Commonwealth Office, and his replacement Sir Kim Darroch, another career diplomat who most recently has been the UK Permanent Representative at the European Union. The NSA also chairs a parallel officials' committee to the NSC, which comprises the civil servant heads of the government departments represented at the NSC.

The government has also set up a National Security Risk Assessment Process (NSRA) to look to potential risks and threats within a five- to twenty-year timescale, whereas those within less than a five-year timescale are prioritized within the latest version of the National Security Strategy (NSS), and the mechanisms of managing them are outlined in the Strategic Defence and Security Review (SDSR).[30] Both the NSS and SDSR are owned by the Cabinet Office, which works to the NSC. So far the main criticism of this new system is that the Cabinet Office lacks the capacity to provide the analytical support that the NSC and NSRA require, and the government has been reluctant to provide additional resources here when it is making significant cuts across government.

For many years, no government articulated national security. Instead, there were the occasional documents produced by the Foreign and Commonwealth Office and the annual Statements on the Defence Estimates until 1999. However, the annual defense white papers ceased and were replaced by the occasional defense white paper, when major changes to defense policy were envisaged. After the U.S. 9/11 attacks, the Labour government increasingly began to speak of the need for a national security strategy modeled on that of the United States. In 2002, it produced a more localized counterterrorism strategy and revised defense white paper.[31] The national security strategy was delayed to March 2008, some six and a half years after 9/11, six years after the publication of 'SDR: New Chapter," five years after the invasion of Iraq, and two and a half years after the attacks on the London transport system, Prime Minister Gordon Brown launched Britain's first National Security Strategy stating:[32]

> . . . the strategy published today will be backed up by *a new approach to engage and inform the public.*[33]

In reality the first NSS provided little more than a potential list of threats and disasters that might befall the UK, with no attempt made to prioritize them. Moreover, there was little substantive policy to deal with the challenges identified. As a result, within fifteen months a second NSS was quietly released.[34] Although the second NSS was again produced by the Cabinet Office, this time it was undertaken with the engagement of the other departments of state, with the result that it tied more closely to their policies. The document repeated the process of providing a list of potential challenges for the UK and again failed to prioritize them. However, this list went further than before, embracing some of the human security literature and thus setting the role of state as not only protecting the state and its citizens but adding the responsibility to protect the citizen from his or her fears. How this was to be achieved was unclear, as policy documents provided any guidance in directing or assisting in the implementation of policy.

As soon as the coalition government was formed in May 2010 it began a process of revising the NSS and undertaking a defense and security review (the SDSR). Both documents were published within a day of one another after some five months of work and immediately before the Spending Review.[35] For the Coalition government, there was a real concern that the United Kingdom's existing credit rating might fall along the lines of the Greek and Irish economies, resulting in the cost of borrowing increasing exponentially. This therefore became the main focus for the coalition government, with the goal of reducing the gap between government income and expenditure so that the financial markets would retain their confidence in the United Kingdom and the credit rating would therefore remain unchanged. Why the new government was surprised by the state of the government's finances is a little surprising. A year earlier David Cameron had already made a speech on the Age of Austerity and its themes had already been applied to the defense context by a number of academic studies.[36] The vision set out in the new National Security Strategy and the SDSR has been coherently constructed, and the two documents dovetail neatly into one another. This is hardly surprising, given that they were originally going to be published as a single document up until a few days before their eventual release, when it was decided to separate them.[37] In fact, the basic tenets of maintaining Britain's place in the world are all too familiar. The coalition has maintained the Blair agenda of continued active engagement in a globalized world and with it the assumptions of a liberal interventionalist agenda, albeit with a far greater emphasis on the national interest and an implied reluctance to become embroiled in further operations, at least in the short term.[38] That said, the first opportunity for the new coalition government to say no produced quite the opposite effect, with British Prime Minister David Cameron joining French President Niklas Sarkozy in leading the international response to Libya.

Nevertheless, the formal approach taken to national security has moved on from merely listing the potential dangers and risks to one in which there is an attempt to evaluate each in terms of likelihood and impact. This has allowed the government to list these various concerns under different tiers in its NSS while its SDSR then categorizes the government's response. This is a distinct move away from previous threat and capabilities-based approaches, and its relative success or failure is only likely to be seen over time. Consequently, it represents a major step forward in conceptual thinking and helps the government justify its defense cutbacks based on the assumption that the United UK's credit worthiness represents its center of gravity.

Reflections

Under the current Conservative-Liberal Democrat government, the UK has taken a series of constructive steps forward in developing the relevant infrastructure to consider the issue of national security and ensure that a whole government approach is taken. Its NSS and SDSR have been the subject of considerable criticism, mainly as a result of the level of defense cuts that are currently being put through the Ministry of Defence and Foreign and Commonwealth Office. Such reductions have a strategic rationale, given the current UK financial situation. However, if this rationale is to be maintained, then the logic of the government's involvement in operation such as NATO operations over Libya remains open to question.

However, two basic challenges remain. First there is the question of whether the union itself will remain or will the Scottish people vote in 2014 for independence. If they do, then the first priority of preserving the union will have failed. Second, there is the debate about what or who constitutes the UK. The UK remains a state that is uncomfortable with itself and has no sense of direction. As a state, it knows what it was, but there is no clear agreed idea of what it wants to become. Thus national security remains the preservation of the UK's current position at a time when the world around it changes. Only once these two issues are resolved can a forward-looking national security strategy really be constructed.

Notes

1. Brian Harrison, *Seeking a Role: the United Kingdom 1951–70* (Clarendon Press: Oxford, 2009); Andrew Marr, *A History of Modern Britain* (London: Pan Books, 2007); George L Bernstein, *The Myth of Decline: The Rise of Britain since 1945* (London: Pimlico, 2004), Andy Beckett, *When the lights Went Out: What Really Happened to Britain in the Seventies* (London: Faber & Faber Ltd., 2009); David Reynolds, *Britannia Overruled: British Policy and World Power in the 20th Century* (Harlow: Longman Group Ltd., 1991); David McCourt, "Rethinking Britain's *Role in the World* for a New Decade: The Limits of Discursive Therapy and the Promise of Field Theory," *British Journal of Politics and International Relations*, vol. 13, no. 2, May 2011, pp. 145–164.

2. See Andrew M. Dorman, *Blair's Successful War: British Military Intervention in Sierra Leone*, (Farnham: Ashgate, 2009).

3. Mike Smith, *Fighting for Ireland? The Military Strategy of the Irish Republican Movement* (London: Routledge, 1995); Paul Bew et al., *Northern Ireland 1991–2001: Political Forces and Social Classes* (London: Sherif, 2002); Michael Cox et al., *A Farewell to Arms? From Long War to Long Peace in Northern Ireland* (Manchester, UK: Manchester University Press, 2000).

4. Michael Cunningham, *British Government Policy in Northern Ireland 1969–2000* (Manchester, UK: Manchester University Press, 2001); Anthony McIntyre, *Good Friday: The Death of Irish Republicanism* (New York: Ausuho Press, 2008); Nick van der Bijl, *Operation Banner: The British Army in Northern Ireland* (Barnsley: Pen & Sword Books Ltd., 2009).

5. Henry McDonald, "Belfast Peace Protestors Rally Against Union Flag Violence," *The Guardian Online*, 13 January 2013, http://www.guardian.co.uk/uk/2013/jan/13/belfast-peace-protesters-union-flag-violence accessed 27 February 2013.

6. "Elect a Champion," *SNP Election Manifesto 2010*.

7. The front page had the title "It'll Cost You" and renamed a series of Scottish places with the implication that independence would cost Scotland dearly. Thus Edinburgh became Edinborrow, while the Outer Hebrides became Outer Cash. *The Economist*, 14–20, April 2012.

8. Simon Johnson, "Sir Richard Dannatt Urges Alex Salmond to Be 'Honest' over Defence," *Daily Telegraph Online*, 18 October 2011; Arabella Thorp and Gavin Thompson, "Scotland, Independence and the EU," *Standard Note 6,110*, House of Commons Library, 8 November 2011.

9. Auslan Cramb, "Scotland 'would have to re-negotiate membership of the EU' in the Event of Independence," *Daily Telegraph Online*, 11 February 2013, http://www.telegraph.co.uk/news/9862464/Scotland-would-have-to-re-negotiate-membership-of-the-EU-in-the-event-of-independence.html accessed 12 February 2013.

10. Andrew M Dorman, "Reorganising the Infantry: Drivers of Change and What This Tells Us about the State of the Defence Debate Today," *British Journal of Politics and International Relations*, vol. 8, no.4, 2006, pp. 489–502.

11. "North-east Votes 'No' to Assembly," *BBC Online*, 5 November 2004, http://news.bbc.co.uk/1/hi/uk_politics/3984387.stm Accessed 14 November 2011.

12. "Boundary Changes: How Could They Affect the UK?," *The Guardian Online*, http://www.guardian.co.uk/politics/datablog/2011/jun/06/boundary-change-constituency-lewis-baston# accessed 14 November 2011.

13. "UK Ethnic Minority Numbers to Rise to 20% by 2051," *BBC Online*, 13 July 2010, http://www.bbc.co.uk/news/10607480 accessed 14 November 2011.

14. "Births in England and Wales by Parents' Country of Birth for 2010," Office of National Statistics, 25 August 2011, http://www.ons.gov.uk/ons/rel/vsob1/parents—country-of-birth—england-and-wales/2010/births-in-england-and-wales-by-parents—country-of-birth—2010.html accessed 24 April 2012.

15. See http://www.london.gov.uk/black-asian-and-ethnic-minorities accessed 12 February 2012.

16. Paul Kennedy, *The Rise and Fall of Great Powers* (New York: Vintage Books, 1987).

17. See Tony Blair, "Doctrine of the International Community," address to the Economic Club of Chicago, 24 April 1999; Tony Blair, *A Journey* (London: Hutchinson, 2010), pp. 2, 47–50.

18. David Cameron, speech to the UN General Assembly, New York, 22 September 2011, http://www.newstatesman.com/node/40963/ accessed 1 May 2012.

19. Jim Murphy, "Preventive Intervention: How the United Kingdom Responds to Extremism in North and West Africa," speech to the Henry Jackson Society, 14 February 2014, http://www.labour.org.uk/how-the-uk-responds-to-extremism-in-north-west-africa-and-beyond,2013-02-14 accessed 27 February 2013.

20. See Michael Howard, *The Continental Commitment: The Dilemma of British Defence Policy in the Era of Two World Wars* (London: Temple Smith, 1972); Basil Liddell Hart, *The British Way in Warfare* (London: Penguin, 1942); C. J. Bartlett, *Defence and Diplomacy: Britain and the Great Powers, 1815–1914* (Manchester, UK: Manchester University Press, 1993).

21. "The Future of the United Kingdom's Nuclear Deterrent," *Cm.6,994* (London: TSO, 2006), http://www.mod.uk/NR/rdonlyres/AC00DD79-76D6-4FE3-91A1-6A56B03C092F/0/DefenceWhitePaper2006_Cm6994.pdf accessed 1 May 2006.

22. S. J. Croft and P. Williams, "The United Kingdom," in *Security without Nuclear Weapons? Different Perspectives on National Security*, Regina Cowen Karp, ed. (Oxford: Oxford University Press, 1991), p. 147; see also Michael Dockrill, *British Defence Policy since 1945* (Oxford: Basil Blackwell Ltd., 1988); William Jackson, *Britain's Defence Dilemma: An Inside View* (London: B.T. Batsford Ltd., 1990); John Baylis (ed.), *Britain's Defence Policy in a Changing World* (London: Croom Helm Ltd., 1977); Ritchie Ovendale, *British Defence Policy since 1945* (Manchester, UK: Manchester University Press, 1994); Stuart Croft (ed.), *British Security Policy: the Thatcher Years and the End of the Cold War* (London: Harper Collins Academic, 1991); Peter Byrd (ed.), *British Defence Policy: Thatcher and Beyond* (Hemel Hempstead: Philip Allen, 1991); Malcolm McIntosh, *Managing Britain's Defence* (London: Macmillan Academic & Professional Ltd., 1990).

23. "Globalization and the Changing UK Economy" (London: Department of Business, Enterprise and Regulatory Reform, 2008), http://www.bis.gov.uk/files/file44332.pdf accessed 1 May 2012, p. 14.

24. See Anthony Seldon and Guy Lodge, *Brown at 10* (London: Biteback Publishing, 2010).

25. "National Population Projections, 2010-based Reference Volume: Series PP2" (London, Office of National Statistics, 29 March 2012), http://www.ons.gov.uk/ons/dcp171776_253890.pdf accessed 1 May 2012, p.1.

26. Erik Goldstein and John Maurer (eds.), *The Washington Conference, 1921–1922: Naval Rivalry, East Asian Stability, and the Road to Pearl Harbor* (Ilford, UK: Frank Cass & Co. Ltd., 1994).

27. See Andrew Dorman, *Blair's Successful War: British Military Intervention in Sierra Leone* (Farnham: Ashgate, 2009).

28. See Frank Ledwidge, *Losing Small Wars: British Military Failure in Iraq and Afghanistan* (New Haven, Conn.: Yale University Press, 2011); Richard North, *Ministry of Defeat: The British War in Iraq 2003–2009* (London: Continuum UK, 2009).

29. See Paul N. Cornish and Andrew M Dorman, "Blair's Wars and Brown's Budgets: From Strategic Defence Review to Strategic Decay in Less than a Decade," *International Affairs*, vol. 85, no. 2, March 2009, pp. 247–261; Hew Strachan, "The Strategic Gap in British Defence Policy," *Survival*, vol. 51, no. 4, August–September 2009, pp. 49–70.

30. "A Strong Britain in an Age of Uncertainty: The National Security Strategy," *Cm.7,953* (London: Cabinet Office, 2010), http://www.direct.gov.uk/prod_consum_dg/groups/dg_digitalassets/@dg/@en/documents/digitalasset/dg_191639.pdf?CID=PDF&PLA=furl&CRE=nationalsecuritystrategy "Securing Britain in an Age of Uncertainty: The Strategic Defence and Security Review," *Cm.7,948* (London: Cabinet Office, 2010), http://www.direct.gov.uk/prod_consum_dg/groups/dg_digitalassets/@dg/@en/documents/digitalasset/dg_191634.pdf both accessed 1 May 2012.

31. "SDR: A New Chapter" (London: TSO, 2002), http://www.mod.uk/NR/rdonlyres/79542E9C-1104-4AFA-9A4D-8520F35C5C93/0/sdr_a_new_chapter_cm5566_vol1.pdf and http://www.mod.uk/NR/rdonlyres/DD89DBE6-CEAA-4995-9E01-52EF6D19FC73/0/sdr_a_new_chapter_cm5566_vol2.pdf accessed 22 March 2008.

32. http://interactive.cabinetoffice.gov.uk/documents/security/national_security_strategy.pdf accessed 21 March 2008.

33. Bold and italics by author, not in original. http://www.number10.gov.uk/output/Page15102.asp accessed 21 March 2008.

34. "The National Security Strategy of the United Kingdom: 2009 Update—Security for the Next Generation," *Cm. 7,590* (London: TSO, 2009), http://webarchive.nationalarchives.gov.uk/+/http://www.cabinetoffice.gov.uk/media/216734/nss2009v2.pdf accessed 1 May 2012.

35. "A Strong Britain in an Age of Uncertainty: The National Security Strategy," *Cm. 7,953* (London: Cabinet Office, 2010); "Securing Britain in an Age of Uncertainty: The Strategic Defence and Security Review," *Cm. 7,948* (London: Cabinet Office, 2010); Andrew M. Dorman, "Making 2 + 2 = 5: The 2010 Strategic Defence and Security Review," *Defense and Security Analysis*, vol. 27, no. 1, March 2011, pp. 77–87; Paul N. Cornish and Andrew M. Dorman, "Dr Fox and the Philosopher's Stone: The Alchemy of National Defence in the Age of Austerity," *International Affairs*, vol. 87, no. 2, March 2011, pp. 335–353.

36. David Cameron, "Age of Austerity," speech made to the Conservative Spring Forum, Cheltenham, 26 April 2009, http://www.conservatives.com/News/Speeches/2009/04/The_age_of_austerity_speech_to_the_2009_Spring_Forum.aspx accessed 18 November 2010; Paul N. Cornish and Andrew M. Dorman, National Defence in the Age of Austerity," *International Affairs*, vol. 85, no. 4, July 2009, pp.

733–753; Paul N. Cornish and Andrew M. Dorman, "Blair's Wars and Brown's Budgets: From Strategic Defence Review to Strategic Decay in Less than a Decade," *International Affairs*, vol. 85, no. 2, March 2009, pp. 247–261;

37. "A Strong Britain in an Age of Uncertainty—The National Security Strategy," *Cm. 7953* (London: TSO, 2010), http://www.direct.gov.uk/prod_consum_dg/groups/dg_digitalassets/@dg/@en/documents/digitalasset/dg_191639.pdf?CID=PDF&PLA=furl&CRE=nationalsecuritystrategy accessed 19 November 2010.

38. See William Hague, "International Security in a Networked World," speech made at Georgetown University, Washington, D.C., 17 November 2010, http://www.fco.gov.uk/en/news/latest-news/?view=Speech&id=117662682 accessed 20 November 2010.

PART III

20TH CENTURY WORLD

6 Australian National Security: The Problem of Priorities

Maryanne Kelton

Introduction

The 21st century challenge for Australia in its national security approach is to manage a bifurcated threat environment that comprises both new and old fears. Though the new security implications of asymmetric threat, climate change, fragile states, cyber security, and resource scarcity have expanded the horizons of security policymakers, reasserting itself as the primary focus of Australian national security strategy is an increasing concern with regional state-based threats. For a period after 9/11, the national security approach became pervaded by the response to terrorism. However, as the 9/11 and terrorist bombings in Bali in 2002 and 2005, where 92 Australians died, have become more distant, the more they have been superseded by a return to state-based threat perception. Recently, the shifts in Asian regional power dynamics have become more prominent in Australian national security thinking. Paramount in the challenge then is to set priorities and to identify the ways and means by which to proceed.

Theoretically, in the design of national security strategy, the state endeavors to protect whatever it considers to be of most worth. Normatively, this calculation includes the security and prosperity of its citizens, but as Melvyn Leffler argues, this assessment also includes the protection of the nation's core values.[1] This understanding is important because of the interconnections drawn between the domestic environment and threat perception. Moreover, the plasticity of this national security approach, "by relating perceptions of threat to core values, helps explain why particular tactics are adopted as policies and others rejected."[2] To understand how the national security approach in Australia is constructed, it is imperative to comprehend how threat perception in Australia is generated. This chapter deploys an integrated threat perception framework as explanatory of the motivations for Australia's national security strategy. It argues that Australian perceptions of threat arise from integrated material and cultural rationales and comprise historic, geographic, and political elements.[3] Historically, Australia's threat perception was initially informed by its existence as a British colony. As past relations are a determinant of threat perception, these social structures became regenerative. Materially but culturally too, after the demise of the British Empire, Australia chose to ally itself with another of the British settler colonial powers, the United States. Not only was it a material superpower, and Australian defense forces have traditionally preferred to operate within larger structures, but as Michael Barnett has argued, the ally decision-making process was shaped by shared norms of democracy and state actor identity.[4] Australia then has a unique position internationally in its relationship with the United States because of British influences in the construction of both states but also because of formal security and intelligence-gathering relationships through its ANZUS, UKUSA, and Defence Trade Treaties, which are complemented by its Australia–U.S. Free Trade Agreement (AUSFTA).

The recent construction of national security strategy in Australia has been shaped by a dynamic assessment of both state and nonstate threats, which have become more visibly expansive this century. The Australian Defence Force (ADF) has been deployed with U.S. forces in the Iraq and Afghanistan construction of the war on terror, and much has been done in Australia at the national government level to reconsider the state apparatus in responding to the breadth of 21st century threats posed. Contemporaneously, the government explained that human security concerns were also catalysts to the offshore actions. Partner to these assessments have been the transnational threats that have asserted themselves forthrightly in the latter part of the 20th century, some of which have been given impetus from the nature of the increasingly globalized world. For example, Australia's environment is sensitive to the manifestations of climate change; it exists in a region where some of the fragile Pacific states struggle to survive, where resource scarcity is on the rise, the health pandemics of SARS and bird flu are prevalent and where WMD proliferation remains a valid threat. As with most states of the globe, it is not immune to the shadowy world of organized crime, including the trade in human migration and also in narcotics.

However, as evidenced in Australia's first National Security Statement (NSS) released in December 2008, the country has become increasingly concerned with executing a policy to ensure its national security by strategic management of its relations with the two key powers of the region, namely the United States and China.[5] As such, this dilemma resonates on the historical divergence of the United States as Australian's military and cultural ally, with China as a primary trade and economic partner with alternate policies on governance, human rights, and freedoms. It does so too in a post-Global Financial Crisis (GFC) environment, which has re-weighted the respective U.S. and Chinese positions and where it is contextualized by the enormous and wider regional demographic growth.[6] For Australia, this growth is occurring in an Asian region that is conceptualized by both "a promise and a threat" and where identity politics still exerts influence on the nature of the Australian security approach.[7] Consequently, in the investigation of Australia's approach to national security, this chapter considers Australia's core interests, its position in the world, and the threat dynamics associated with that position. The chapter then analyzes how Australia has responded to these threat perceptions, both in the construction and maintenance of its security apparatus and in its alliances and security partnerships.

Australian Perceptions of Its Changing World Position

Undeniably, Australia's relative position in the world is a fortunate one. It is a stable, liberal capitalist democracy in the southwest Pacific. It is part of a Western alliance underpinned by U.S. strength and membership of the ANZUS security alliance and UKUSA intelligence-sharing treaties. Reflecting the manifestations of Australia's prosperity, in health, education, income, environment, and equity, the 2010 UN Development Programme Report ranked Australia second of all countries listed on its global human development index.[8] It is the 14th largest economy globally and 9th largest industrial economy. Notwithstanding its vulnerability to flooding rains, drought, and the ravages of bushfire, Australia remains home not only to approximately 21 million people, including those from two hundred countries, but also to 10 percent of the world's biodiversity, together with significant deposits of natural resources, including coal and iron ore.[9]

As a consequence of its prosperity, political and social stability, educated workforce, and natural resources, to date it has possessed some of the strategic assets to maintain an attractive quality of life for many of its citizens and claim a status as a midrange power. This position, however, is not without constraint and limitation, nor is it free from both inflated threat perception on the one hand and complacency on the other. Australia's purchase on the maintenance of this position and its ability to shape international affairs to its advantage is limited by the structural constraints of a small population, its position as a middle ranking power in the international system, and the relative growth of its regional neighbors, and as always is subject to the capacities of its political leadership. Moreover, Australia's political will to act is at times captive to the dilemmas arising from the disparate nature of its Anglo history and its Asian geography and, correspondingly, the degree to which it can proceed beyond its domestic incarceration within its historical legacy.

In 2008, Australia asserted that it was "a regional power, prosecuting global interests."[10] As both a proponent of middle power diplomacy, particularly when the Australian Labor Party retains power, and a proponent of bilateralism when the Coalition Party (Liberal and National Parties) holds office, the state regards itself as a key regional player. That Australia is a developed country in a developing region certainly has given some credibility to its regional claims. That it has been a co-innovator in the establishment of liberal institutions such as Asia Pacific Economic Cooperation (APEC) in 1989 and the ASEAN Regional Forum (ARF) in 1994, in addition to its leadership of the UN intervention teams such as the International Force for East Timor (INTERFET) in the 1999 East Timor sovereignty crisis and the 2003 Regional Assistance Mission Solomon Islands (RAMSI) lend some authority to its self-declared regional and local neighborhood importance. Materially too, the military strength of its high-tech defense forces that have boasted regional air and naval superiority in the Southeast Asian region also impart support to the government's position. Australia's aspirations, however, are extending more widely with its candidacy for a nonpermanent seat on the UN Security Council (UNSC) in 2013–2014. It is campaigning under the banner of "making a difference for the small and medium countries" and on the strength of a sixty-five-year commitment to the UN; its UN security and financial contributions; its multilateral activism in arms control and trade; and its promise to double its aid budget to AUD8 billion by 2015.[11]

However, Australia's assessment of its global security interests and its capacity to prosecute them has been much shaped by its U.S. alliance. Indeed, it is hard to imagine that without the U.S. impetus, Australia would have deployed forces to Iraq in 2003. Additionally, it was U.S. diplomatic, economic, and military leverage exercised over Indonesia that allowed for the Australian-led UN Force to enter East Timor in September 1999. And it was U.S. logistic support that assisted Australia to transport its forces to East Timor during this period. Australia's claims to military strength stem partially from its partnership within the U.S. alliance, yet the costs in blood and treasure of deployment in problematic interventions remain evident. In some regions of the globe these deployments with the Anglosphere detract from UN members' support of Australia's candidacy for the UNSC seat.

Australia's economic clout has been enhanced by its resilience during the 2008–2009 GFC, and its relatively more buoyant economic record has been modestly impressive in comparison with other developed economies. The number of new employment positions in 2010 was the highest ever recorded, and most of the 490,000 new jobs created since mid-2009 were full-time positions. Overall unemployment in December 2010, which was 5.1 percent,[12] compared favorably with the 10 percent figures sustained in the EU and the United States. Its economic legitimacy was acknowledged in its membership of the recently convened G20 forum. Yet at the same time, Australia's immunity to the depredation of the GFC has been derived from the luxury of its trading relations with China. In December 2010, Australia recorded a AUD$14 billion trade surplus with China on the back of a robust raw materials trade.[13] So herein lies the paradoxical security dilemma for Australia: as tensions rise between the world's two major powers, the United States and China, Australia's capacity, already limited, to craft a policy managing its relations with these major powers becomes increasingly complex.

Nonetheless, the paradox of regional change and the consequence for national security questions for Australia are more substantive than just the China question. Australia's location astride the South Pacific and Indian Oceans, once derided because of the tyranny of distance to the European metropole, is progressively regarded as advantageous as it is more proximal to the eastward shift of centers of global economic power. Yet the combination of the rapidly expanding demographics with the return of economic growth to the region particularly post-Asian financial crisis of 1997–1998 suggests that Australia will eventually be dwarfed by its neighbors. By 2050, ten of the nineteen states estimated to have populations over 100 million will be located in Australia's region of primary security concern. In addition to China, these states include India, the United States, Indonesia, Japan, Pakistan, Bangladesh, the Philippines, Vietnam, and Russia. Of these, five are nuclear weapons states, with Japan a potential nuclear state and North Korea threatening its recently acquired nuclear status provocatively in the sustenance of its regime. Though the strength of regional states will limit China's potential to influence, it also adds to the complexity of the regional power dynamics and relationships with which Australia must engage. Furthermore, Australia must attend and adapt to the dynamism of this regional environment, where the Asian continent has substantially increased trade and investment flows, and which generates and oversees energy flows from west to east.[14] As Michael Wesley has subsequently argued, of proximal importance then to Australia is Southeast Asia's location at the potential intersection of Great Power competition in this burgeoning new Indo-Pacific geostrategic realm.[15] His argument posits that one of the greatest challenges for Australia is to overcome the complacency of its current strong economic position and the previously relatively benign security status of its immediate environs.[16] Long-standing Australian security analyst Coral Bell contends that the most significant challenge to Australia will be how it responds to the dynamics of these security challenges, in all their manifestations, with the demand for the resources of energy, water, and capital being unprecedented.[17] Through the Australian National Security Statement (NSS) of 2008, former Prime Minister Kevin Rudd extrapolated these security implications to East Asian and Southwest Pacific threats, including those of intrastate conflict, transnational crime, people smuggling, health pandemic, and climate change. Similarly, the legacy of the terrorist threat continued to attract considerable attention in the NSS.[18]

The uncertainty of the post–Cold War world for Australia is gradually being replaced, but only by the increasing certainty of change in the regional dynamics. On the one hand changes will occur in state-based power sharing, and on the other, likely increases in transnational threats. The Australian domestic response to these developments remains mixed as a consequence of the divergent pattern of its international economic and security relationships. Though many Australians benefit from the economic relationship with China and many businesses seek to maximize their profitability through interconnections into the Chinese marketplace, Australia's traditional security arrangements and dominant cultural legacy continues to inform the way threats are perceived. Interestingly, for a period at the end of the conservative government incumbency in the mid-2000s, after Prime Minister John Howard had significantly promoted the prominence of the U.S. relationship, the political space was created for a more sanguine perception of China within the domestic sphere. However, the consequences of an inflation of tensions in the Sino-U.S. relationships will reverberate in Australian popular perceptions. And as Leffler has argued, it is these perceptions that also inform the approach to national security.

National Security Articulation, Organization, and the Recent Evolution of the Security Apparatus

Australian national security issues are articulated via the Defence White Papers and more recently through the NSS. The latter marks the culmination of change over the past decade and as such it is one of three periods of reform of the national security apparatus in Australia.[19] Post 9/11, in the review of internal and border security and its coordination, former Department of Defence Secretary Ric Smith advised against the establishment of a Department of Homeland Security. In his judgement, a super department model risked less accountability and flexibility. Potentially it could also focus inward and function counterproductively to the assessment across the suite of possible threats, which for Smith included everything from terrorism, energy management, organized crime, cyber attacks, and border security to coastguard formation.[20] Coherent coordination instead was sought through alternative means and directed initially by Rudd's NSS. Instead of a new homeland security bureaucracy, the government chose to bolster existing capacities and coordination via the establishment of specific roles and committees. The PM's opening gambit in 2008 was to appoint former SAS Commander Duncan Lewis as the National Security Advisor (NSA), with responsibilities to coordinate the government activity, set the strategic policy framework, and oversee the budgets. The new NSA would also chair the National Intelligence Coordination Committee (NICC) and act as deputy-chair of the Secretaries Committee. Responsive to the political sensitivities of the public and media regarding human smuggling, the NSA would also chair the Border Protection Taskforce. Beyond these structures, Rudd also established the Counterterrorism Control Centre, National Crisis Coordination Committee, and Criminal Intelligence Fusion Centre. Additionally, an Office of the Australian Civilian Corps was initiated within AusAID to implement the initiative, and a National Security College with funding of AUD$17.3 million over three years was opened.[21] Impressive as this list was long on new administrative measures, one experienced analyst wryly observed that "if committees can make us safe, then we're going to be a very secure country indeed."[22]

In sum, the NSS was "roundly criticized as a missed opportunity that lacked detail and gave little guidance to security agencies in dealing with ever-growing threats to the country."[23] Though its advocacy of a risk-based approach was commendable, it was unable to provide detail as to how this approach would be prosecuted. And though it identified a list of priorities by which the national security interests would be fostered, these were so broad as to mostly only identify all means by which Australia has proceeded to address national security previously. For example, five of the priorities were to develop or strengthen ADF capabilities; the U.S. alliance; regional relations; nuclear disarmament; and Pacific states' economic health. To this broad and weighty list was added the implementation of the Smith Report on homeland and border security; enhancing e-security; and incorporating climate change and energy security concerns into the approach.[24] By the time of this writing, its desire to foster an Asia-Pacific Community had been rendered unworkable.

It is hardly surprising in this new impetus for national security that government expenditure in the sector grew exponentially. The total of government spending on these new national security initiatives in the decade to 2010 will approximate AUD$10 billion, a sum that did not include defense expenditure.[25] By way of a specific example, the Australian Secret Intelligence Organisation (ASIO) experienced an increase in funding of 539 percent, from AUD$69 million in 2001–2002 to AUD$441 million in 2008.[26]

Partner to these organizational changes in the responsiveness to threat, however, lies a parallel consideration. How does the state manage the threat responsiveness so that it maintains the respect for the values of civil liberties upon which the Australian state was constructed? When a state assiduously pursues the denial of threat realization and acts to protect itself from danger, the state must ensure that it does not compromise its core values. Limits to the range of the NSS exist. As Michael Hogan argued of U.S. national security measures, "[t)]he American people and their leaders, or at least the best of them, would go so far and no further, lest a reckless abandon destroy the very republic they sought to protect."[27] In the years immediately post-9/11 and the Bali bombings, the Australian Law Society questioned the implementation of "draconian and disproportionate" legislative infringements on the rights and freedoms of Australian citizens.[28]

The "National Interest"

In Australia, the phrase "the national interest" has suffered from overuse in the political lexicon and has been deployed politically to justify the political parties' rationale for their respective national security policies. While both John Howard's Coalition government (1996–2007) and more recently the Rudd and Gillard Labor governments have determined ultimately that the national interest is the preservation of Australia's security and prosperity, its liberal capitalist disposition, and its rule of law and civil liberties, the extension of its use has been as a rationale for the method by which the preservation of security in its broadest sense is approached. The phrase was brought to prominence in the public domain in 1997 when the conservative Howard government produced two Foreign Affairs and Trade White Papers successively entitled *In the National Interest*[29] and *Advancing the National Interest*[30] designed to articulate how DFAT policy purpose was to advance "The security and prosperity of Australia and Australians," which prioritized political and economic freedom and valued "tolerance, perseverance and mateship."[31] The government also eschewed the previous Labor government's prioritization of multilateralism and regionalism for "practical bilateralism," which reinstituted a public focus on traditional alliances. Threats and responses were perceived through the prism of Australia's vulnerabilities, its past relations, and its inherited identity. It was also responsive to the electorate's lethargy for persistent change in its engagement with Asia as driven by the previous Hawke-Keating Labor governments.[32] After 2001, the Conservative government also utilized the phrase to justify their military activism in Iraq and Afghanistan. They did so on the basis that Australian security was best defended by supporting U.S. objectives in denying security threats before they reached either Australian shores or its immediate region.

Change arrived, however, when Labor governments from 2007 reinstituted the national interest concern for security conflicts closer to home and in the preparation for high-end wars. In announcing the production of the 2009 Defence White Paper, former Labor Defence Minister Joel Fitzgibbon declared the government's commitment to "the defence of Australia, the security and stability of the regional security environment, and a rules based global security order."[33] While it withdrew combat troops from Iraq, it increased its commitment to Afghanistan. If the region warranted greater attention because of shifting power dynamics and concerns for fragile states in the Pacific, the legacy of Australia's participation against a nominated jihadist threat, its involvement in Afghanistan, and the broad bipartisan commitment to the U.S. alliance remained intact. Even though the Labor government prioritized the immediate region in its pronouncements, it nevertheless attempted to cover the multiplicity of Australia's concerns. It argued that the approach to national security must deal with broad challenges in delivering "a stronger Australia given the long term challenges to our economy; a fairer Australia given the levels of disadvantage that continue to exist among us; and an Australia capable of meeting the sweeping new challenges of the 21st century, including climate change."[34]

Noteworthy in this new rhetorical conceptualization of national security was the downgrading of the terrorist threat and its replacement with elevated climate change concerns. Prior to 2007, Australia had failed to ratify the Kyoto Protocol despite its negotiation of diminished arrangements within its Annex 1 country categorization. However, accompanying Labor's election in 2007 was the fanfare surrounding its preparedness to respond more expeditiously to climate change. With its eventual ratification of the Kyoto Protocol in Bali at the UN Framework Convention on Climate Change, the government publicly announced its sensitivity to the 21st century variations in national security demands. Yet despite its assiduous preparations for the Copenhagen Climate Change Conference in 2009, the collapse of these talks punctuated the government's ability to prosecute its climate change policy domestically. Locally too, public support for climate change action in Australia without corresponding international action was not sustainable. Domestic intransigence in some sections of Australian society to climate change policy and a pricing of carbon remains durable with Prime Minister Gillard's attempts to enact change floundering. That Australia has a well-developed and highly profitable coal industry that supplies approximately 80 percent of its energy needs and contributes substantively to its export surpluses adds to the difficulties in climate change policy formulation. For Australia then the development of national security policy regarding climate change stumbles at the gap that exists between its middle power activism and the domestic obduracy to proceed in any kind of first-mover role.[35]

Partners and Allies

One of Australia's fears as the Cold War was drawing to a close in the early 1990s and as the Uruguay trade round extended was the possibility it may be isolated in a world where broad trade blocs competed internationally. As the previous discussion emphasizes, Australia by dint of its location encountered an unusual predicament. The crucial question for Australia remained that of the identity of its natural partners. New Zealand is a natural fit and a partner in assisting Pacific Island concerns, but size beleaguers its potency internationally. As a British settler colonial state, Canada exhibited some of the same characteristics as Australia, but its U.S. land border altered both the military and financial security dynamics of its international motivations. In the Anglo "coalition of the willing," the UK had previously divested itself of more intimate connections to its former colony in the mid-1960s when it relinquished its security connections East of Suez and joined the European Economic Community in 1973. So it was for Australia to make its way in Asia, where it was finding ready trade partners. Yet Japan's WW2 aggression and Australia's cultural apprehensions made security partnerships problematic for some time. Nevertheless, as eventual spokes in the U.S. system of Pacific alliances, and indeed as the north and south anchors of U.S. security strategy in the Pacific, the democracy of Japan offered some opportunity for regional diplomacy. In the arena of multilateralism, Australia and Japan's work on the International Commission on Nuclear Non-proliferation and Disarmament remains as evidence of their developing cooperative activism. However, in 2007 Australia eschewed the opportunity to participate in quadrilateral talks with the United States, Japan, and India, given China's *demarche* against the grouping. The country not much discussed in the Australian literature is Indonesia. Though Australia and Indonesia have worked cooperatively to respond to some of the regional people smuggling concerns in the past few years, this pairing has yet to realize its potential.

President Obama may have categorized Australia with a suite of others when he stated that "*the United States does not have a closer or a better ally than Australia,*"[36] but undeniably, the United States remains as Australia's preeminent security partner and the backbone of Australia's regional security approach. Its relationship with the United States was the central motivating feature of Australia's post 9/11 deployments. Nevertheless, in the immediate post–Cold War period, Australia also sought to situate itself as a mature state cognizant of the role that regional institutional building and evolving bilateral relations would have in the sustenance of Australian security. Spurred by successive Prime Ministers Hawke and Keating, Australia was influential in the construction of the ARF (despite U.S. initial objections) as a means by which East Asian security institutional building could develop. Comprising the ASEAN states plus states with security interests in the region including Australia, Canada, China, India, Japan, New Zealand, Russia, South Korea, and the United States, the forum was designed to contribute to preventive diplomacy through dialogue, transparency, and confidence-building. Though the ASEAN Defence Ministers Ministerial (ADMM+) has evolved from this forum, Australia's preference is for both institutions to more actively address security concerns. ASEAN's disinclination to pragmatically address the East Timor crisis of 1999 was demonstrative of this reluctance. If state-based intervention is problematic, there is no question that Australia has an interest in this group functioning, at the least, to address the suite of regional transnational concerns that include securing the maritime environment and preventing the spread of communicable disease. As Australia attempts to confront the interconnected security problems of fragile Pacific Island states as they experience the problems of poor economies of scale and governance in improving the social and economic circumstances for their populations, it looks to New Zealand for support. However, critiques of Australian diplomacy oscillate between that of declining influence and accusations of interventionist hegemony. As illustrative of the diminution of Australia's South Pacific influence, General Bainimarama, self-installed leader after the latest coup in Fiji, has recently assumed a position as head of the Melanesian Spearhead Group contrary to Australia's requests. Noteworthy here is that China provides funding for the Melanesian Spearhead Group. Moreover, China continues to provide extensive financial support for Fiji, as its economy has contracted dramatically since Bainimarama's usurpation of power.

Reactions to Systemic Change

Though new security issues have come to prominence at the outset of this century, systemic change is repositioning the state to the forefront of the national security questions for Australia. Accompanying the stated expenditure for the expansion in the national security apparatus are ongoing defense spending increases. As a response to the expanded nature of threat perception the government has committed to a real increase of 3 percent in defense spending to 2017–2018 and to 2 percent to 2030.[37] In the May 2010 budget, a total of AUD$25.7 billion was allocated for defense spending.[38] The 2009 Defence White Paper reflected the renewed Australian focus on protecting its national interests in an era of dynamism in the geostrategic landscape and with the possibility of great power rivalry. All of which were exacerbated by the varying effects of the GFC and the ongoing need for stability in the global economy. As former Defence Minister Joel Fitzgibbon observed, the government's assessment of systemic change is that Australia can expect

> changed strategic power relativities and an increasingly 'multipolar' global order driven by changing patterns of underlying economic power and political influence.[39]
>
> [Further] there are likely to be tensions between the major powers of the region where the interests of the United States, China, Japan, India, and Russia intersect. As other powers rise, and the primacy of the US is likely to be tested, power relations will inevitably change'.[40]

In making the judgment that the United States has a range of global interests and may be stretched elsewhere, the government is of the opinion that Australian assistance, presumably to assist in convergent objectives, may be required in regional security matters.[41] Alternately, the Rudd government argued that China's responsibilities to explicitly document its strategic intentions as an adjunct to its increasing power projection capabilities were manifest.[42]

While the Labor Party's framework for advancing Australia's interests internationally rests on the three pillars of commitment to the U.S. alliance, regional engagement, and multilateral commitment, principally through the UN and its agencies, Australia's response to these changes to date has been to express its commitment to working with the United States and to advocate strengthened regional institutions that are inclusive of the United States in the maintenance of regional stability and order. At the 2010 Australia U.S. Defence and Foreign Ministerial meetings (AUSMIN), both states declared their plan to further develop naval, space and cyberspace ties, notable as these are areas in which China has escalating capabilities.[43] Though undeniably China must address successfully a suite of internal social and economic challenges[44] and manage the current fracturing of its foreign policy voice, it provides significant financial input into the Australian economy through its purchase of raw materials. Correspondingly, Hillary Clinton's AUSMIN talks question, "how do you toughly deal with your banker?,"[45] highlighted her concerns regarding Australia's divergent loyalties.

As Australians themselves began to grapple seriously with the likelihood of regional changes and how economic strength backed by an extended demographic could impact upon regional power, the debate was inevitably precipitated as to how Australia *should* react to the shifting ground. It is understood by some that while the United States has global interests, Australia's interests are essentially regional and not always convergent with those of the United States. Also evident in Australian–U.S. relations has been a naive willingness of Australian policy-makers to rely on cultural dimensions of the U.S. relationship to generate influence and deliver outcomes.[46] If former Prime Minister Malcolm Fraser reiterated his view that a "slavish devotion" in responding to U.S. military objectives was antithetical to Australia's long-term security interests,[47] former Deputy Defence Secretary Hugh White urged Australians to reject the underlying assumption that Asian strategic power and order will remain unaffected by its economic revolution.[48] An inevitable consequence of China's likely prosperity was its intersection with U.S. East Asian power arrangements. As such, Australian complacency needed to be replaced by a shrewd, pragmatic, and immediate reassessment of its options in this context where power politics have been reinstated and a construction of a regional defense grouping with its Southeast Asia neighbors may ultimately be preferable over an undiscriminating reliance on the U.S. alliance; the transfer to a new ally; a position of armed neutrality; or as is occurring at the moment, a sanguine lethargy to think, choose, and act to limit the dangers of its future choices.[49] Australia's attempt to have it all was being seriously eroded by the intersection of changes in the international system and the traditional inclinations of the Australian populace. It is also evident that the outcomes of these power arrangements not only decide the harmony of the region between states per se but will ultimately determine the efficacy of the multilateral responses to the transnational threats, which include the capacity to respond to climate change.

Reflections

Since the demise of the Cold War, the processes of globalization have increased the intimacy of new security risks and threats to which the Australian government has increasingly chosen to respond. Offshore, Australia's initial military deployment to Afghanistan post–9/11 and then Iraq was analogous with its historical imperial defense strategies, albeit in this era, a response to a variation on traditional security threats. Australia promptly and assuredly acted to shore up its U.S. alliance credentials and join its partners, the United States and the UK. Correspondingly, onshore, in its national security approach the obsession with counterterrorism drove the way in which national security was conceived and restructured. It was also true that threat amplification served the Coalition government's electoral claims. By doing so it also played to the political right's promotion of the intersection of the terrorist threat with asylum seeker immigration.

Undeniably, an integrated and cohesive government approach is likely to be more effective in the response to the threats presented at the outset of the 21st century, particularly given the increasing blurring of the international and domestic dimensions of threat. The extraordinary funding increases in Australia since 2001 certainly have added capacity to the ability to capture and analyze information. The appointment of a national security advisor provided much-needed oversight to the coordination of the approach. Yet these measures should not be construed as a fundamental restructuring of the approach to national security but as capacity and coordination-building measures, particularly with reference to the new security threats. Moreover, effectiveness in performance, and not just compliance, the recording of indicators, and the language of outcomes, is required. In an era too where the corporatist tendency has been to eschew specialists, the reinstitution of the value of longitudinal specialist knowledge should also assume greater importance.

Ultimately, the extraordinary spending on security apparatus and defense procurement will need to demonstrate its effectiveness, particularly as Australia's outlook increasingly returns to its regional sensitivities, aware of the shift in regional state-based power dynamics. Consequently, as the security space has expanded, the overarching policy imperative for Australia is to determine its priorities and how it will act upon them given the limitations of its resources. As Eric Morse argued in his critique of the Obama national security strategy, "if everything is a priority, then nothing is."[50] Currently in Australia it is unclear not only as to the country's priorities for national security but also how it will resolve this issue. To date, it has been inescapably observable that Australia's alliance with the United States has steered the direction of the national security approach. As the United States now seeks to draw down its deployments in Iraq and Afghanistan, Australia should reacquaint itself with not only East Asian security politics but also those of the Indo-Pacific. Australia as the southern anchor of its Pacific alliance system presents itself as an attractive U.S. ally with which to conduct the burden sharing of the defense of the Western Alliance. While Australia shares a similar values system with the United States, the problem for Australian policy-makers is to calculate what this greater alliance burden sharing may entail and to comprehend that cultural attachments alone are no substitute for rational material assessments of Australian interests or capacity to exercise influence. As two regional analysts argued in 2001, "as a security policy for Australia (acting within the U.S. framework), the deepest flaw in this approach remains what it has always been: U.S. and Australian interests are not always convergent and are not made more convergent merely by Australian fidelity."[51] The irony of Australia's national security approach is that Australia's exports to China enable the expansion of the security apparatus and its defense procurements. The question remains as to the length of time these exports can underwrite Australia's national security strategy.

Notes

1. Melvyn P. Leffler, "National Security," in *Explaining the History of American Foreign Relations*, 2nd ed., Michael J. Hogan and Thomas G. Paterson, eds. (Cambridge: Cambridge University Press, 2004), pp. 123–125.

2. Ibid., pp. 131–132.

3. See Maryanne Kelton, *More than an Ally: Contemporary Australia US Relations* (Aldershot: Ashgate, 2004). Klaus Knorr, *Historical Dimensions of National Security Problems* (University of Kansas, 1976), p. 78; Graeme Cheeseman, "Australia: The White Experience of Fear and Dependence," in Ken Booth and Russell Trood (eds.), *Strategic Cultures in the Asia-Pacific Region* (Basingstoke, Macmillan, 1999, p. 79); Alexander Wendt, "Constructing International Politics," *International Security*, vol. 20, no. 1, Summer 1995, p. 74.

4. Michael Barnett, "Identity and Alliances in the Middle East" in *The Culture of National Security: Norms and Identity in World Politics*, Peter Katzenstein, ed. (New York: Columbia University Press, 1996), pp. 400–447.

5. Kevin Rudd, "The First National Security Statement to the Parliament," address by the prime minister of Australia, Canberra, 4 December 2008.

6. Coral Bell, "Living with Giants: Finding Australia's Place in a More Complex World," ASPI, Canberra, April 2005.

7. Juanita Elias and Carol Johnson, "On Re-engaging Asia," *Australian Journal of Political Science*, vol. 45, no. 1, March 2010, p. 9.

8. UNDP, "International Human Development Index: Australia," http://hdrstats.undp.org/en/countries/profiles/AUS.html, accessed 14 January 2010.

9. DFAT, "Australia in brief," 2011, http://www.dfat.gov.au/aib/overview.html.

10. Rudd, 2008.

11. DFAT, "Australia: Candidate for the United Nations Security Council 2013–14," Canberra, 2011, http://www.dfat.gov.au/un/making_a_difference.html.

12. Australian Bureau of Statistics (2011), "6202.0 Labour Force Australia December 2010," 13 January 2011, Canberra, http://www.abs.gov.au/ausstats/abs@.nsf/mf/6202.0.

13. DFAT, "China," December 2010, http://www.dfat.gov.au/geo/fs/chin.pdf.

14. Anthony Bubalo and Malcolm Cook, "Horizontal Asia," *The American Interest*, May-June 2010.

15. Michael Wesley, "SEA-Blindness, Why Southeast Asia Matters," East-West Center, Washington, March 2011.

16. Michael Wesley, "There Goes the Neighbourhood: Australia and the Rise of China," UNSW Press, 2011.

17. Coral Bell, 2008.

18. Kevin Rudd, 2008; and Kevin Rudd, Speech at the Opening of the National Security College of Australia, Canberra, 24 April, 2010.

19. The first of these in the 1970s was responsive to the Hope Royal Commissions and mandated the creation of the Office of National Assessments to funnel intelligence assessments to the executive of government. The second of the reforms occurred post–Cold War in response to a prospective peace dividend, and the third in the early 2000s. See Carl Ungerer, 2010.

20. Rudd, 2008.

21. Kevin Rudd, 2010.

22. Hugh White, "Analyst Gives PM Mixed Report Card," interview, The World Today, ABC Radio, 4 December 2008, http://www.abc.net.au/worldtoday/content/2008/s2437751.htm.

23. Richard Pearlman and Cynthia Banham, "PM Security Statement 'a damp squib,'" *Sydney Morning Herald*, 5 December 2008.

24. Rudd, 2008.

25. Carl Ungerer, "Connecting the Docs: Towards an Integrated National Security Statement," *Australian Strategic Policy Institute,* 10 December 2009, p. 2.

26. Carl Ungerer, "The Intelligence Reform Agenda: What Next?," *Australian Strategic Policy Institute*, 27 February 2008.

27. Michael J. Hogan, *A Cross of Iron: Harry S Truman and the Origins of the National Security State, 1945–1954* (New York: Cambridge University Press, 1998), pp. 474–475, 485.

28. Law Council of Australia, "Legal Profession Opposes Anti-Terror Bill," Media Release, 2 November 2005, and Law Council of Australia, "Summary Comment on the Anti-Terrorism Bill (No. 2)," 14 November, 2005.

29. Commonwealth of Australia, *In the National Interest* (Canberra: 1997).

30. Commonwealth of Australia, *Advancing the National Interest* (Canberra: 2003).

31. Ibid., p. vii.

32. Maryanne Kelton, 2004.

33. Joel Fitzgibbon, "The 2009 Defence White Paper: The Most Comprehensive White Paper of the Modern Era," Canberra, 2 May 2009, http://www.defence.gov.au/whitepaper/mr/01_OverarchingWhitePaperMediaRelease.pdf.

34. Rudd, 2008.

35. Malcolm Cook, "The Problems of Political Sustainability," paper delivered at the Flinders University in association with the Chinese Academy of Social Sciences Developing Sustainable Societies conference, Adelaide, 22–23 March 2011.

36. Barack Obama quoted in Transcript of Joint Press Conference with the President of the United States of America, Yokohama, 13 November 2010, http://www.pm.gov.au/press-office/joint-press-conference-yokohama.

37. Rudd, Speech to the National Security College, Canberra, 24 April 2010.

38. John Faulkner, Minister for Defence, May 2010, http://www.defence.gov.au/minister/Faulknertpl.cfm?CurrentId=10273.

39. Commonwealth of Australia," Defending Australia in the Asia Pacific Century: Force 2030," Department of Defence White Paper, Canberra, 2009, http://www.defence.gov.au/whitepaper/docs/defence_white_paper_2009.pdf, p. 30.

40. Ibid., p. 33.

41. Ibid., p. 32.

42. Ibid., p. 34.

43. AUSMIN, Joint Communique, Canberra, 8 November 2010, http://foreignminister.gov.au/releases/2010/kr_mr_101108.html.

44. Nouriel Roubini, "China Guesses Wrong on Growth," *Australian Financial Review*, 18 April 2011.

45. "U.S. Embassy Cables: Hillary Clinton Ponders US Relationship with Its Chinese 'Banker,'" *The Guardian*, 4 December 2010.

46. Kelton, 2004.

47. Malcolm Fraser, "Slavish Devotion to the US a Foreign Policy Folly for Australia," *The Age*, 14 December 2010.

48. Hugh White, "Powershift: Australia's Future between Washington and Beijing," *Quarterly Essay*, 39, 2010.

49. Idem.

50. Eric S. Morse, "Analysis of the Obama Administration's National Security Strategy 2010," *The National Strategy Forum Review*, 3 June 2010, p. 2.

51. Wade Huntley and Peter Hayes, "East Timor and Asian Security," in *Bitter Flowers, Sweet Flowers: East Timor, Indonesia and the World Community*, Richard Tanter, Mark Sheldon, and Stephen R. Shalom, eds. (Lanham, Md.: Rowman and Littlefield, 2001), p. 181.

7 Providing for National Security: Canada After 9/11

David Rudd[1]

Introduction

In May 2011, Canadians elected their first majority government in six years. During the election campaign the contestants said little about national or international security, despite the country being embroiled in two major military campaigns. The rejection of Canada's bid for a seat on the UN Security Council (UNSC) the previous year—a first for a country that sees itself as one of the world body's greatest boosters—likewise made no ripple on the domestic political waters. Nor did the party leaders articulate their visions for the evolution of the security relationship with the United States. Indeed, there was scant evidence that any of the major parties attached great importance to the whole question of security or Canada's place in the world.

This may be no more than a brief snapshot of the Canadian political landscape—a testament to the old adage that all politics is local, at least at election time. But it is at odds with the reality that Canada has made significant efforts to come to grips with security challenges on the home front and in a rapidly changing international system. While the political foundations upon which the country's security culture and foreign policy have remained largely the same over time, the resources and institutions devoted to their pursuit have undergone a noticeable transformation since the epochal events of 9/11.

This chapter will assess Canada's position in the global community and explore the style and substance of Canada's approach to national security, how it articulates and carries out policy, and how it may be positioning itself to deal with some of the issues and trends that characterize the evolving domestic and international order. The picture that emerges is one of a country that is at once fortunate in its geopolitical position and vigorously engaged in the business of promoting security at home and abroad. At the same time it is preoccupied with a range of challenges, including in its northern regions. The extent to which the muscular rhetoric of the Harper government heralds the emergence of a new foreign policy "doctrine" will also be explored.

Canada's Place in the World: Continentalism versus Internationalism?

With the world's second-largest landmass, longest coastline, abundant natural resources, proximity to the world's largest economy, and membership in some of the world's most exclusive clubs, Canada is in a most advantageous geopolitical position. Flush with hosting the 2010 Winter Olympics (at which it set a record for gold medals won) and the heads of state at the G-8/G-20 meetings, and having conspicuously avoided the deepest traumas of the global financial crisis, the country has displayed a degree of confidence and international visibility that ordinarily escapes it, being so close to its culturally similar yet infinitely more influential neighbor to the south.

In 2011, Canada's GDP growth had slowed but still surpassed $1.7 trillion, ranking it eighth in the OECD. Its population density was the 12th lowest in the world, and the number of foreign-born citizens exceeded 20 percent of a population of 34 million. This suggests that Canada is an attractive (and secure) place to settle—a sentiment reflected in research by the UN Development Program, which ranked Canada as the sixth most livable country in the world in its quality of life index.[2] Despite having almost 3,000 military members deployed on UN-sanctioned operations in early 2012, the Global Peace Index found Canada to be the fourth most peaceful nation out of 158 surveyed.[3]

Canadians have traditionally viewed themselves as global players with privileged access to the halls of power but without much of the ideological baggage that characterizes Great Power relations. Those professionals charged with representing them on the international stage have been more pragmatic, knowing that Canada is an actor of decidedly modest means. Accordingly, the country has long styled itself as a "middle power"—one of among many Western liberal-democratic states for whom acting in concert with like-minded partners is a more pragmatic strategy than trying to exercise decisive leadership on global issues. The term was coined by Canadian diplomats in 1942[4] and has been a fixture of Canadian foreign policy ever since, alternately used to distinguish Canada's international stature from larger powers (i.e., the permanent members of the UNSC) and to foster a sense of nationhood at home.[5]

Canada also prides itself on its commitment to pluralism, multiculturalism, and respect for the rule of law and for civil liberties. Though this does not set it far apart from other Western states, it allows Canadians to think of themselves as creators of a polity that serves as an example for others to follow.[6] It should not come as a surprise, therefore, that a harmonious, consensus-seeking domestic political culture should be reflected in Canada's dealings with the outside world. There it adheres to the tenets of liberal internationalism: the promotion of a rules-based order and a prominent role for institutions in regulating interstate behavior and promoting peace and security. Canada was present at the founding of the UN and retains membership in several regional security/discussion bodies—notably NATO, the Organization for Security and Cooperation in Europe, and the Organization of American States.

But while it has often been an enthusiastic supporter of global governance initiatives, such as the International Criminal Court (ICC), and has sought to liberalize norms of security behavior—such as those embodied in *The Responsibility to Protect* (R2P)[7]—it has occasionally gone outside the formal institutional system in order to advance important causes. It joined with like-minded governments and nongovernmental organizations to spearhead the campaign against antipersonnel landmines in the mid-1990s.[8] In 1999 it participated in the NATO-led air war over Kosovo in the absence of explicit UN Security Council authorization. Thus, while committed to formal multilateralism, Canada has shown a willingness act through informal "coalitions of the willing" in the broadest sense of the term.

Ottawa has also failed to meet the UN goal of devoting 0.7 percent of GDP to ODA, and the government of Stephen Harper (in power since 2006) has drawn criticism for refusal to ratify the Kyoto Protocols on climate change. There is little evidence to suggest that the government opposes foreign aid or environmental protection *per se*, but one may surmise that it has concluded that the opportunity cost of implementing these policies was too high, or provided insufficient net benefit to Canada's security—at least in the short term.[9]

Bilateral ties with the United States continue to be the essential ingredient in Canada's international engagements, with economic and security cooperation dominating the agenda. Cross-border trade totaled US$596 billion in 2011[10]—the largest bilateral trade relationship by volume in the world. Canada is also the largest foreign supplier of energy to the U.S. market, exporting crude oil, natural gas, uranium, and hydroelectric power that equals 9 percent of America's total demand.[11] Canada holds a highly privileged position in the U.S. national security apparatus by virtue of its sharing of responsibility for the defense of continental aerospace. Despite significant differences in resources between the two partners, NORAD is a partnership of equals wherein a Royal Canadian Air Force officer may be entrusted with the power to assess whether the continent is under attack and recommend a response to both national command authorities. Further trust between the two neighbors is manifested by the oft-quoted fact that the two neighbors share the world's longest undefended border (approximately 8,900 km).

But the depth and breadth of security cooperation has changed since the 9/11 attacks, with the border having been reinforced by joint patrols along key waterways involving the U.S. Coast Guard and marine units of the RCMP. Flights of American unmanned air vehicles along the border between the province of Manitoba and the state of North Dakota began in 2009 as a way of combating drug smuggling. They have since been expanded from Washington State eastward to Minnesota.[12] Canada has also agreed to supply the Department of Homeland Security with data from twenty-two radar feeds to help fill gaps in the Federal Aviation Administration's surveillance of the airspace straddling the two countries.[13]

Canada's perennial search for—and what some would consider its preoccupation with—recognition from its southern neighbor begs the question as to the degree to which its security behavior is too Continentalist, or, in the words of one analyst, designed primarily "to secure a favourable place in the American sphere."[14] Leaving aside the fact that ODA (a key component of the overall security-building effort) is not targeted at U.S. recipients, or that trade-dependent neighbors are compelled to take border security seriously, there is scant evidence that Canada's recent bout of risk-taking in Afghanistan or Libya (what one may call robust multilateralism) has translated into tangible benefits in other policy areas. There is no expectation of American benevolence on trade disputes, the resolution of which are in any case influenced by constellations of industrial and congressional interests.[15]

Nevertheless, critics have observed a contraction in some of Canada's international engagements and have painted a picture of a once-active international citizen in decline.[16] Some commentators point to the marked decline in contributions to UN peacekeeping missions since the end of the various Balkan crises as evidence that Canada's "traditional" role of "helpful fixer" acquired over several decades has been undermined. Such criticisms are usually accompanied by harsh words for Canada's support for NATO's mission in Afghanistan,[17] but the data underpinning the critique is accurate. The number of Canadians wearing blue berets on *traditional* peacekeeping missions has shrunk to barely two hundred from the low thousands at the end of the 1990s. However, when one considers that peacekeeping is now understood to encompass a wide range of operations along the conflict spectrum, the picture is quite different. One study noted that personnel devoted to peace support operations have consistently been in the 2,500–3,000 range since 9/11; the main difference being that Canadian troops now tend to be deployed on missions that are UN approved but not UN led, and that they are concentrated in a smaller number of missions and geographical areas, of which South Asia has recently been the most prominent.[18] If factoring in the cumulative contributions of Canadian personnel—both military and civilian—to the various relief missions in Haiti (culminating in the 2010 earthquake disaster),[19] it is difficult to argue that any alleged abandonment of UN operations has resulted in a net loss to regional or international security.

There are also other ways of measuring Canada's commitment to peace and security beyond the mere presence of soldiers. By supplying equipment and training to foreign forces—the latest example being the UN mission to Darfur—Canada has empowered missions to which it otherwise makes no human resource contributions. Increasingly, police officers and officials from the Department of Foreign Affairs and the CIDA are on call to serve overseas. Foreign Affairs' START was established in 2005 to aid in the reconstitution of government services in Afghanistan. Since 2006, Canada has allocated $450 million to the Global Peace and Security Fund to support reconstruction and peace-building in the world's worst trouble spots. On the policy development side, Canada has enthusiastically supported efforts to curb atrocities in conflict zones through R2P (ironically, it is NATO that is enforcing these norms contained in Libya). Efforts to build a consensus in the UN to protect civilians and aid groups threatened by war have also met with success.[20] Thus, while troops still count for much, ideas, leadership, development funds, and civilian expertise are highly prized and may, from the government's perspective, constitute a worthwhile contribution when military resources are tied up elsewhere.

Periodically, Canadians might, to their great disappointment, misperceive their place in the international order. Such disappointment was briefly on display in October 2010 when the holy grail of Canadian internationalism—a nonpermanent seat on the UNSC—was denied to Canada. Although it had successfully lobbied for a seat in each of the ten-year intervals in which it had applied for candidacy, Canada lost to tiny, nearly bankrupt Portugal. Veteran foreign policy scholar Denis Stairs summed up the reaction of the attentive public thus:

> Canadians are accustomed to thinking of their country as a darling of the UN: supportive of it from the start; perpetually attentive to its purposes and programmes; a pre-eminent participant in peace operations; a dedicated believer in multilateralism, along with the development of international institutions and the rule of international law; the home of a foreign service particularly adept at diplomatic conflict resolution and of a military establishment universally renowned for its empathetic peacekeeping skills on the ground; and all the rest. They may, of course, be quite wrong in believing all this and in presuming that others think of them in the same way as they think of themselves. But right or wrong, they tended to interpret the outcome of the vote as a stern rebuke, and to wonder what their country had done to deserve it.[21]

Stairs points to the increase in the number of UN member-states eligible for a seat and the tendency for countries to vote for candidates who hail from their particular region as possible reasons for the setback in Canadian fortunes. Seen from this perspective, Canada had little control over the strength of its bid; the law of averages would reduce the likelihood of anyone winning, and if Europe wished to have a greater voice in the Security Council, the choice of EU member Portugal was entirely logical. In any case, secret balloting gives no hint as to why those who initially supported Canada's bid changed their minds.

That the story of Canada's failure quickly faded from the headlines undoubtedly came as a relief to the government. Sadly, this may have removed any incentive for sober reflection as to what, if anything, Canada might do to (re)gain a place at one of the world's most prestigious tables. A temporary place on the Council affords middle powers the opportunity to, however modestly, influence the international security agenda, thereby promoting and protecting their own security.

Canada's National Interests: Definition, Determinants, and Evolution

Since the immediate postwar era, successive governments have sought to buttress Canadian security along three lines: the protection of the homeland, the defense of the continent in cooperation with the United States, and contributing to international security alongside allies and partners. Although the priority given to each of these has changed slightly over time, with defense and security budgets fluctuating accordingly, these three macro-tasks have remained largely intact. Cooperating with the United States in matters relating to territorial defense and the policing of the border to ensure the free flow of trade have historically been Canada's main security goals. (The latter suggests an awareness that national security encompasses far more than the mere defense of the homeland against internal and external armed threats but extends to the nation's economic health.)

Canadian foreign policy went through a brief period in the 1990s in which human security and the exercise of soft power were given (rhetorical) primacy, resulting in support for a number of international initiatives that were, admittedly, outward focused; Canadians would be indirect beneficiaries of the R2P or other measures aimed at promoting good governance abroad. By strengthening the international system, Canada would, theoretically, be promoting security at home. But concerns with national security, broadly defined, seemed to take on a greater degree of importance during the tenure of Liberal Prime Minister Jean Chrétien, who placed heavy emphasis on expanding trade links with emerging economies. Efficacy aside, the many Team Canada trade missions led by the prime minister himself during the 1990s spoke to the government's belief that economic prosperity was a key component of national well-being.[22]

The terror attacks of 9/11 reordered the priorities again, or rather united them into a more integrated whole. Protection of the homeland and citizenry was still the primary concern, although new issues had been "securitized." The adversary was now no longer a foreign power or constellation of powers wielding military might; it was now a nonstate actor, ensconced in regions of political upheaval (failed/failing states), and looking to benefit from the proliferation of WMD. Like the Cold War era, Canada would have to take measures to reassure its allies (especially the United States) that it would not allow its territory to be used as a base or conduit for attacks on others. But unlike the previous era, policies and resources would be focused, at least initially, on the nonmilitary arms of the national security apparatus.

Articulating National Security Policy

The first major policy initiative of the new era was the Anti-Terrorism Act. Promulgated in late 2001, it provided new guidance to law enforcement agencies and allowed Canada to ratify various international conventions designed to suppress terrorist financing, thwart bombings, and safeguard UN and associated personnel.[23]

Intent on keeping the Canada–U.S. border open for trade, the Chrétien government also concluded a series of technical agreements with Washington under the Smart Border Declaration and Action Plan. The plan's main objectives included the secure flow of people and goods, investments in secure infrastructure, and a pledge to share information and coordinate activities in the pursuit of these objectives. The main federal agencies involved in this whole of government endeavor included Canada Border Services (see the following section), the RCMP, and Foreign Affairs.

By 2004, the Liberal government of Paul Martin was in office and determined to codify Canada's approach to the new security environment. It quickly published *Securing an Open Society: Canada's National Security Policy*,[24] which restated the three main security goals (macro-tasks) and proceeded to list a broad range of issue-areas that would be addressed. These included improved (domestic) intelligence collection, enhanced border and transport security, and improved emergency management capabilities across the various federal departments (yet another example of what would become the whole of government approach to security). Interestingly, the document specified that it would enhance capacity to address public health emergencies—a reference to the outbreak of SARS in the spring of 2003—illustrating the expansive definition of national security. It also took pains to point out that activities would be carried out in a manner consistent with the tenets of liberal democracy, confirming that the preservation of core values was also a matter of security.

The government also pledged to address what it perceived to be Canada's inability to shoulder its fair share of the international security burden. In

2005, the IPS was released. Entitled *A Role of Pride and Influence in the World*, the IPS acknowledged that the security environment was threat-ambiguous but sought to quantify (mainly) indirect threats to national security. These included failed and failing states, terrorism, and the proliferation of unconventional weapons. Although Canada might not be the target of these threats, their ability to threaten core interests (i.e., U.S.–Canada trade) and values (i.e., the safety of civilians) was highlighted. To address this complex reality, the government decreed that an integrated "3-D approach" comprising diplomacy, defence, and development (or combinations/permutations thereof) would henceforth be adopted.[25]

Protecting Canada and Canadians remained the top job in the IPS's subordinate document, the DPS. Additional resources were promised for an expansion of the military's Special Forces capability, as well as measures to defend against attacks by unconventional weapons. Intelligence spending was given a boost, and the military was ordered to liaise more closely with civil authorities—federal, provincial, and municipal—who might share overlapping responsibilities for domestic security. The government also specified that greater efforts should be made to secure Canada's vast northern regions.[26] The salience of failed/failing states gave impetus to efforts to improve certain expeditionary capabilities that had atrophied following the budget cuts of the 1990s.

The momentum continued after Stephen Harper's Conservative government ascended to power in 2006. The 2008 *Canada First Defence Strategy* was not fundamentally different from the IPS/DPS in terms of strategic outlook. Rather, it took a more realistic view of the cost of rebuilding the Canadian Forces and pledged additional funds, including those required to increase Canada's military presence in the Arctic.[27] Three years later, Harper unveiled the *Northern Strategy*, laying out the government's intent to protect Canadian sovereignty, safeguard the northern environment, and promote socioeconomic development for indigenous peoples.[28] It compliments Canada's membership in the Arctic Council, an eight-nation grouping established prior to 9/11 but which has assumed greater prominence as concerns over climate change and northern resource development have been added to the security agenda.[29] Another example of Canada's long commitment to multilateralism, Ottawa has nevertheless been reluctant to grant observer status in the Council to several European states and to the EU as a whole. One may infer that Canada's commitment to multilateralism has limits; it will not allow its influence in key policy forums to be diluted by what it considers to be outsiders—no matter how friendly or well meaning.[30]

The National Security Apparatus Since 9/11

At the federal level, Canada has long maintained a wide array of agencies that contribute in their respective ways to national security. The principal ones include the following:

- The Royal Canadian Mounted Police (RCMP)—law enforcement (also under contract to provide services to provincial, territorial governments, and some municipalities)
- Canada Border Services Agency (CBSA)—customs, border security (created in 2003 under Public Safety Canada)
- The Canadian Security Intelligence Service (CSIS)—counterintelligence
- Communications and Security Establishment Canada—signals intelligence support and protection of electronic information and communication
- The Department of National Defence and the Canadian Forces (DND/CF)

Other agencies that could be considered to have a security role include the following:

- Citizenship and Immigration Canada—screens immigrants, refugees, foreign students and temporary workers
- Canadian Coast Guard—a special operating agency of Fisheries and Oceans Canada sharing search and rescue duties with the CF

Since the attacks of 9/11 two new organizations have been created:

- Public Safety Canada (PSC)—interdepartmental security coordination
- The Canadian Air Transport Security Authority (CATSA)—passenger/luggage screening

PSC is perhaps the most significant domestic organization to take shape since 9/11. Established in 2003, it has a dual mandate: to work with other government departments and operators of critical infrastructure (i.e., hospitals, utility companies) to help ensure essential services in the event of an emergency, and to run the Government Operations Centre, which monitors threats to the national interest on a twenty-four-hour basis. One of its interdepartmental tasks is to work alongside Citizenship and Immigration to issue security certificates, which can be issued to noncitizens who are in Canada illegally and/or are judged as posing a serious threat to Canada.

Another actor that has taken the stage in recent years is the national security advisor, who acts as a senior advisor to the prime minister in matters relating to security and intelligence. The post was created by the 2005 national security policy and provides a single point of contact for the production and coordination of intelligence, the assessment of which is then passed to senior government officials. Among the departments feeding information into that pool, CSIS is perhaps the best known and most open to public scrutiny. It gathers information, at home and abroad, on major threats such as terrorism, WMD, and foreign espionage. Its budget has almost doubled since 2011.[31]

The most recent (and little-known) change has been the establishment of the Cabinet Committee on National Security to "provide broad strategic direction for security and foreign policy related to Canada's national interest."[32] Whereas theses tasks once fell under the purview of the Foreign Affairs and Defence Committee headed by the defense minister, the new body will be chaired by the prime minister himself, indicating that security issues will henceforth receive greater attention.

Since 2005, DND/CF has received significant injections of financial resources to meet new and enduring security challenges. While defense spending as a percentage of GDP has remained relatively stable, expenditures have risen from C$13 billion in 2001 to C$23 billion in 2010.[33] This can partially be accounted for by the trebling of spending on peace support operations during that time.[34] But recruiting and capital acquisitions are also benefiting. The size of the regular force has risen from 59,000 to 68,000 in the 2001–2011 time frame.[35] Among the most notable programs to be put forth by the Harper government is the pending acquisition of six to eight ice-capable offshore patrol vessels. Designed to operate in northern waters during the navigable season, they represent a sea change in focus for the Royal Canadian Navy. They will fulfill a purely constabulary role in cooperation with other federal fleets, illustrating the importance attached to northern security and Ottawa's faith in the whole of government approach as a means of achieving it. The establishment of Marine Security Operations Centres (MSOCs) in Halifax, Victoria, and in the Great Lakes region will likewise bring stakeholders from various security agencies together to ensure that Canada's east and west coasts, along with the commercially vital St. Lawrence Seaway, are protected from environmental threats, illegal fishing, and transnational crime.[36]

Another notable change in the maritime domain involves the alteration of the venerable NORAD partnership. In 2006 the responsibility for monitoring U.S. and Canadian maritime approaches and internal waterways was handed to the binational command. However, unlike the aerospace function in which the command could assign assets from either country anywhere within continental airspace, responses to maritime incursions will be the responsibility of the nation in whose waters the event is taking place. This somewhat curious arrangement may be attributed to perennial Canadian sensitivities about being too closely integrated with the U.S. national security apparatus, and the U.S. armed forces in particular.

Systemic Changes and Canada's Response

Given the expansive nature of national security, the economic fragility of Canada's biggest trading partner may have profound and long-term implications for the well-being of Canadians. The generation of revenue through trade flows will influence budgets for the various departments and agencies involved in the security sector. Even if the U.S. government avoids defaulting on its debt, the sheer size of the U.S. fiscal imbalance may result in a lengthy period of economic uncertainty for America's trade partners and debt-holders around the world. According to one analyst: "Canada, as an exporting country doing three-quarters of its trade with the U.S., would share in the pain."[37]

While Ottawa has little influence over how America deals with its fiscal woes, efforts to ensure a secure border will undoubtedly continue. A meeting between Prime Minister Harper and President Obama in February 2011 provided some indication that the Smart Border program may evolve into a broader perimeter security arrangement. In this, the partners would more closely integrate their law enforcement, critical infrastructure protection, and cyber-security efforts in order to "accelerate the legitimate flows of people and goods between both countries, while strengthening security and economic competitiveness."[38]

Public acceptance of greater security integration will have to allay fears over Canadian sovereignty, including the traveler's right to privacy and the autonomy of Canada's immigration and refugee-determination systems. But the very fact that deeper collaboration is being contemplated shows the degree to which Canada must take into account U.S. security concerns if it wishes to ensure the security of the populace. In the meantime, Ottawa is mounting a diplomatic offensive in China to drum up business and is pursuing free trade negotiations with the EU. These are clear signs the trade diversification is considered essential to long-term economic security, as well as a recognition that the global distribution of power is changing.

The preoccupation with complexity in the international system is also spurring reactions in Canada's security thinking. As previously noted, the challenge posed by failed states, terrorism, and other direct and indirect threats have compelled practitioners to eschew purely military responses to external threats and embrace a whole of government approach to crisis management. This has since been superseded by the comprehensive approach, which draws in participants from outside government ministries. The participation of nongovernmental organizations and international bodies in the same mission has been practiced (albeit with limited success) in Afghanistan, Haiti, and elsewhere. The concept has been adopted by the government of Canada and has gained enough traction internationally so as to be prioritized in NATO's 2010 Strategic Concept.[39]

The heightened awareness of complexity is reflected in the preoccupation of national security agencies with the so-called global commons. This refers to the international maritime, air, space, and cyberspace domains over which no national jurisdiction is recognized. Use of these domains for commercial activity and other security-related endeavors by all states and legal nongovernmental entities creates a large stakeholder community wishing to protect it for licit use. Accordingly, the Royal Canadian Navy has contributed to counter-piracy operations in the Indian Ocean, and Public Safety Canada has released a strategy to protect critical computer networks from unauthorized access by foreign intelligence-gatherers or from outright attacks by cyber criminals.[40]

The opening up of the Arctic to greater commercial activity has raised the specter of competing territorial claims by Arctic states—specifically, the extent of maritime boundaries. While all eight countries bordering the Arctic Ocean agree that any such disputes should be settled peaceably using the UN Convention on the Law of the Sea, many are making security investments in order to demonstrate a credible presence in this inhospitable region. Aside from the aforementioned offshore patrol ships, Ottawa intends to recapitalize the Canadian Coast Guard, including building the largest ice-breaker ever to fly the flag. It has also announced plans to construct a northern docking facility to refuel government ships patrolling northern waters. A promised military training facility at Resolute in Nunavut will enhance cooperation between regular military units and the Canadian Rangers, a para-military force composed of indigenous Canadians tasked with patrolling the country's vast north. Since 2007, additional budgetary allocations have increased the Rangers' ranks from 4,100 to 4,700. A target of 5,000 is expected to be met by 2012.[41]

Systemic changes are not confined to the external environment. While political élites usually define the methods and the goals, portraying issues or invoking threats as a truly national matter, domestic constituencies can seek to define security in a manner suited to a different set of outcomes, including highly parochial ones.

Canada's Tamil diaspora—the largest of its kind in the world—has long argued for diplomatic involvement in the civil war in Sri Lanka. It has vigorously (and successfully) lobbied successive Liberal governments to refrain from designating the Liberation Tigers of Tamil Eelam (LTTE) as a terrorist organization, which would have brought the attention of police and security forces to its fund-raising activities in Canada and elsewhere. Upon coming to power in 2006, the Harper government outlawed the LTTE and did not heed calls by the expatriate community to stop the Sri Lankan military from ruthlessly crushing the insurgency.

Though the government may not have perceived a link between the security of a foreign population with relatives in Canada and the national interest, the same might not be said for other diasporas. As noted, Canada has been a fixture in international efforts to relieve suffering in Haiti, with some suggesting that this activism is due in no small measure to the expatriate population living in vote-rich Montreal, as well as the fact that Canada's former governor-general (in office during the 2010 earthquake) is of Haitian ancestry.[42]

Canada's Partners and Allies: Narrowing the Field?

In view of the breadth and depth of the bilateral relationship, Canada–U.S. relations will continue to be the lynchpin of Canada's national security. However, the relationship has the potential to be the most problematic, given the clear asymmetry in economic, military, and, indeed, cultural power.

Canadians will continue to display varying degrees of antipathy toward demands by their neighbor to secure the North American space, fearing that the progressive alignment of security policies will undermine Canadian distinctiveness. Mexico, the other partner in the North American economic space and Canada's third-largest trading partner, does not receive the same consideration as the United States despite the shared interest in thwarting the transnational drug trade. According to one observer, "In its desire to smooth flows at our own border, Ottawa has kept Mexico out of talks with Washington on a North American [security] perimeter."[43]

Beginning in 2009, Canada welcomed Danish air force surveillance aircraft to airfields in northern Canada. This was in recognition of the shared interest in monitoring ship traffic—especially cruise ship traffic—in waters between Greenland and Baffin Island. It also provided insights on the feasibility of jet aircraft operating from gravel runways, of which there are many more than paved ones in the far north.[44] In January 2011, the members of the Arctic Council signed an agreement to divide the region into zones for which each would be responsible for search-and-rescue. This marked the first time that the Council, which had previously been restricted to making recommendations on security, had been used to conclude a binding treaty.

Further afield, Canada's multilateral partnerships seem alive and well, although government officials have criticized NATO allies for their perceived unwillingness to contribute to the mission in Afghanistan. Defense Minister Peter MacKay singled out Germany in 2008, saying that Berlin could do more to counter the Taliban in the volatile south of the country.[45] (At the same time, German and Canadian forces were collaborating to supply the latter with modern Leopard 2 tanks and trained crews for duty in Kandahar province.) But for much of Europe, preoccupied as it is with the debt crisis and the evolution of EU institutions, Canada is a distant security partner, and it is an open question whether a free trade deal will change that. Although Canada has made substantial contributions to high-risk operations in Afghanistan (ground troops) and Libya (strike/reconnaissance aircraft, naval vessels), as well as to the alliance's commonly funded programs,[46] the reflexive ties to Europe are eroding as Canada's ethno-cultural mosaic traces its lineage less and less to the "old world."

That said, there is no indication that Canada will jettison its membership in NATO, if for no other reason than its allies represent a group of familiar, generally like-minded states with capable, albeit shrinking, militaries.

Although some allies have either eschewed active participation in the Libyan operations, or placed caveats on the use of their forces in theater, Canadian officials have expressed no dissatisfaction with the level of cooperation. For their part, the allies appear to have confidence in the appointment of a Canadian air force officer as the mission commander.

But within this large family, there are allies to whom Canada has established a deeper level of cooperation and trust. The armies of America, Britain, Canada, Australia, and New Zealand (known collectively as ABCA) speak to Canada's desire for long-term partnerships based on linguistic and operational interoperability. This pattern is replicated by the partners' navies (minus New Zealand), who participate in annual rim of the Pacific (RIMPAC) exercises. Intelligence-sharing with U.S. and British agencies is routine, although information-gathering by CSIS operatives overseas will may require on-going liaison with other foreign (including nonallied) intelligence bureaus.

Whither the National Security: A "Harper Doctrine"?

Since forming the government in 2006, the Conservatives under Stephen Harper have made the rebuilding of the Canadian Forces a high priority while at the same time cutting back at the Department of Foreign Affairs. The latter has seen its budget pared to $1.9 billion in 2010/11 from $2.6 billion in 2007/08, leading to charges that Canada is de-investing in diplomacy at a time when the international security landscape is becoming more complex.[47]

These measures have been accompanied by a flurry of strong rhetoric from the government on Arctic sovereignty, including public criticism of the U.S. ambassador to Canada on the subject of Washington's nonrecognition of Canada's claim to sovereignty over the Northwest Passage. This was accompanied by strong words for Russia, whose long-range bombers continue to test NORAD's reaction time,[48] although none have actually entered Canadian airspace. In private, the prime minister is reputed to be more pragmatic and measured, going so far as to oppose a greater role for NATO in the Arctic on the grounds that armed conflict in the region is unlikely.[49]

Still, this raises the question of whether there is a campaign to establish a particular national security brand or, as some have ventured to say, a Harper Doctrine. Possible characteristics of the latter include a refocus on relationships with traditional allies (especially the United States), military transformation, free trade, strong assertions of Arctic sovereignty, and, as mentioned, vocal support for embattled democracies such as Israel. The promotion of democracy and human rights would also be part of the equation—goals entirely consistent with Canada's foreign policy tradition. But the means of achieving them would not be confined to holding conferences or rendering development assistance; it would include strong critiques of state behavior. This was the case with China during the early months of the government's mandate, where the prime minister declared that human rights would not be sacrificed on the alter of bilateral trade. It was more recently on display in September 2012 when the government decided to close the Canadian embassy in Tehran and expel Iranian diplomats from their mission in Ottawa.[50]

To the extent that there is a new doctrine taking shape, it may be a matter of style rather than substance. In a revealing interview with *Maclean's Magazine*, Prime Minister Harper alluded to his government's impatience with the value placed on Canada's image as an international "good guy":

> [I]t isn't enough, in this day and age, to say we get along with people. We have to have a clear sense of where we want to be and where we would like our partners to go in the various challenges that are in front of them. Whether they're economic challenges or security challenges or anything else . . . it's not just good enough to say, "everybody likes us."[51]

In terms of substance, Harper's security priorities may point to a desire to chart a new course for Canada. Said one observer: "The Conservatives didn't like the direction in which Foreign Affairs had been running because so much expertise developed over a period of a couple of decades on Africa, on human security, on landmines—good Liberal themes."[52] But on close examination, the security initiatives pursued by the government have differed little from those of its predecessor: a Liberal government sent troops to Kandahar, began the process of rebuilding the CF, and took risky steps keep the Canada–U.S. border open to trade. Moreover, respect shown to the Canadian military transcends the Conservative's base of support, the pursuit of free trade not inconsistent with past practice, nor is the commitment to northern sovereignty apt to find disfavor among the Canadian public. Indeed, there seems to be more continuity than change.

One discernible stylistic change from previous practice has been Harper's rhetoric at the UN, where according to Jordan Smith, he has taken issue with the values and priorities of an organization that most Canadians would have regarded as sacrosanct:

> In [his] first speech to the United Nations in September 2006, [Harper] signalled a dramatic shift by questioning the international body's relevance in language that might just have easily been used by Jeanne Kirkpatrick, President Reagan's ambassador to the UN . . . He criticized the pace of UN reform, and was sceptical of the organization's effectiveness in Haiti, Sudan and Lebanon, and about the new Human Rights Council . . . Harper has since challenged the UN on its perdurable hostility toward Israel.[53]

Whether this style will cause the government to demote the UN in the hierarchy of actors expected to deliver security for Canada and its allies is unclear. As of 2012 the government's priorities have shifted to expanding trade with nontraditional partners in the Pacific Rim, and fostering closer security and technical cooperation with fellow Arctic states.[54] But however the government asserts itself, it will almost certainly have to loosen the political reigns and rely on the expertise of a revamped diplomatic corps—one versed as much in the government's priorities as in the complexities of the international system—if it wishes to navigate its way through the security environment. Summing up the challenge awaiting the government, veteran journalist Brian Stewart opined:

> Any real attempt to promote Canada as purposeful new force in the world will require a full-scale rebuilding of Foreign Affairs to get it back to the point where its views are listened to and its skills are deployed to the fullest. A handful of CF-18 fighters and a frigate won't suffice.[55]

Similarly, the practice of channeling ODA to places where Canada has immediate security concerns (i.e., Afghanistan, Haiti) rather than distributing it in small packets to a broad range of recipients might be entirely rational. But it might also impede other security goals such as arming Canada with the credibility needed to sway a sufficient number of countries to vote for it when nonpermanent seats on the UN Security Council are up for grabs. The reduction and eventual termination of Canada's Afghan mission in 2014 may provide the political space for a review of this approach, and to revisit the salience of the nonmaterial components of national security. In the words of one newspaper editorial:

> We must learn from the Afghanistan experience and be prepared through effective professional development, education and training among all government departments that may contribute to future international comprehensive security operations that will be conducted in Canada's interests.[56]

Reflections

For a country that rarely invokes the term, for a people who do not typically make it an issue at election time, Canada is clearly interested in national security in the broadest sense. A decade's worth of legislation, bureaucratic reforms, relationship-building, military transformation, and so forth point to a recognition that the well-being of the country requires a multipronged approach, even if certain components of the overall security equation (the environment, for instance) are not currently part of the agenda.

How that agenda may change is open to debate, but it may be reasonably assumed that bilateral concerns will compete fiercely with broader international ones for the attention of policy-makers, regardless of who holds the levers of power and however much the public fails to appreciate that they are two sides of the same coin.

But Canada will also continue to seek a place of (relative) prominence on the international stage and will view the security landscape from what it considers to be a unique vantage point because there is considerable domestic support for it. Although the recent bout of muscular internationalism in Afghanistan and Libya may strike some as out of character, Canadians are a generally pragmatic lot ready to contribute to and, occasionally, make sacrifices to establish a stable, liberal international order. While perhaps not as prone to risk-taking as some, they sense that national security is not purely national; it means engaging with rather than isolating themselves from the world and all its challenges. These challenges may compel the current or future government to seek new relationships, reprioritize the main security tasks, or even articulate new ones. Much of this work may go on behind the scenes, with little public knowledge or input. But even if the Canadian people do not appreciate the full breadth of the transformation of the security-building effort, those charged with providing for national security have been well and truly active. By this measure, and if the events of the last decade are any guide, the days of episodic interest in the national security file are over.

Notes

1. The views expressed herein are those of the author and do not necessarily represent those of the government of Canada, the Department of National Defence, or the Canadian Forces.

2. Sarah Magee, ed., *Strategic Profile: Canada 2011*, Toronto: Canadian International Council, 2011, (accessed 30 September 2011 at http://opencanada.org/wp-content/uploads/2012/11/SSWG-Strategic-Profile-2012.pdf.pdf).

3. Institute for Economics and Peace, Global Peace Index 2011 Fact Sheet, 1 (accessed 24 July 2012 at http://www.visionofhumanity.org/wp-content/uploads/2012/06/2012GPI-Fact-Sheet2.pdf).

4. Adam Chapnick, "The Middle Power," *Canadian Foreign Policy*, vol. 7, no. 2 (Winter 1999), p. 74.

5. Tom Keating, "Multilateralism: Past Imperfect, Future Conditional," *Canadian Foreign Policy*, vol. 16, no. 2 (2010), p. 12.

6. Government of Canada, *Canada's International Policy Statement: A Role of Pride and Influence in the World*, 2005, p. 4.

7. The product of the UN-sponsored International Commission on Intervention and State Sovereignty, *The Responsibility to Protect* was informed by the experiences of Canadian Major-General Romeo Dallaire, whose underequipped UN force was prevented in taking proactive measures to head off the Rwandan genocide. It laid out principles governing the circumstances under which the international community might intervene—diplomatically or otherwise—in the affairs of a nation unable or unwilling to protect its population from harm.

8. These initiatives were pursued under the direction of the activist former foreign minister Lloyd Axworthy, who campaigned to place human security—the promotion of the individual rather than the state as the unit of analysis in international relations—at the core of Canadian foreign policy. See Christina Badescu, "Canada's Continuing Engagement with United Nations Peace Operations," *Canadian Foreign Policy*, vol. 16, no. 2 (2010), p. 47.

9. Office of the Prime Minster of Canada, "Prime Minister Stephen Harper Calls for International Consensus on Climate Change," 4 June 2007 (accessed 30 May 2011 at http://pm.gc.ca/eng/media.asp?id=1681). Said Harper to a Berlin audience: "[W]hen we came to office [in 2006], Canada's emissions were 33% above the [Kyoto] target and rising. Which meant, with only months before the targets kicked in, it had become impossible to meet the Kyoto commitment without crippling our economy."

10. U.S. Government, "Trade in Goods with Canada" (accessed 16 June 2012 at http://www.census.gov/foreign-trade/balance/c1220.html). This compares to $527 billion in 2010, $431 billion in 2009, $600 billion in 2008, and $566 billion in 2007.

11. Government of Canada, "Canada-U.S. Energy Relations," (accessed 14 May 2011 at http://www.canadainternational.gc.ca/washington/bilat_can/energy-energie.aspx?lang=eng). These figures are for 2009.

12. Security Technology News, "Unmanned Spyplane for Canadian Border Security" (accessed 15 May 2011 at http://www.security-technologynews.com/news/unmanned-spyplane-for-canadian-border-security.html).

13. Sheldon Alberts, "Canada to Share Border Radar Surveillance with U.S.," *Global News*, 17 May 2011 (accessed 18 May 2011 at http://www.globalnews.ca/Canada+share+border+radar+surveillance+with/4802291/story.html).

14. Keating, p. 20.

15. The twenty-year dispute over tariffs on Canadian softwood lumber exports (imposed when the United States alleged unfair subsidization) was settled in 2006 largely in Canada's favor by North American Free Trade Area and World Trade Organization panels. This was accomplished in the face of strong and persistent challenges by the U.S. lumber industry and its allies in Congress. See CBC News, "Softwood Lumber Dispute," 23 August 2006 (accessed 27 May 2011 at http://www.cbc.ca/news/background/softwood_lumber/).

16. These critiques are based on either a review of government activity and expenditure across a broad range of activity or on specific security files such as international peacekeeping. See Andrew Cohen, *While Canada Slept* (Toronto: McClelland and Stewart, 2004), and A. Walter Dorn, "Canadian Peacekeeping: Proud Tradition, Strong Future?" (accessed 20 July 2011 at http://walterdorn.org/pub/32).

17. Steven Staples, *Marching Orders: How Canada Abandoned Peacekeeping and Why the UN Needs Us Now More Than Ever* (Ottawa: Council of Canadians, 2006).

18. Osvaldo Croci, "Canada: Facing up to Regional Security Challenges," in E. Kirchner and J. Sperling, eds., *National Security Cultures: Patterns of Global Governance* (London: Routledge, 2010, 136–139). The author notes that expenditures on peace support operations climbed steadily from C$266 million in 2000/01 to almost $1.2 billion in 2008/09.

19. Department of National Defence, "Operation HESTIA" (accessed 10 July 2011 at http://www.comfec-cefcom.forces.gc.ca/pa-ap/ops/hestia/index-eng.asp). The web page lists Canada's contributions to Haiti's security under the heading "Previous Canadian operations in Haiti."

20. Badescu, pp. 50–53.

21. Denis Stairs, *Being Rejected in the United Nations: The Causes and Implications of Canada's Failure to Win a Seat in the UN Security Council* (Calgary: Canadian Defence and Foreign Affairs Institute, March, 2011, p. 1).

22. Beginning in 1994, these missions visited China, Russia, and Germany. In addition to taking business leaders along, Prime Minister Chretien was often accompanied by provincial premiers, illustrating the intergovernmental nature of economic security building. In the end, the various missions may have not significantly enhanced trade. See April Fong, "Are Trade Missions a Thing of the Past?" *The Globe and Mail*, 15 October 2010 (accessed 3 July 2011 at http://www.theglobeandmail.com/report-on-business/are-trade-missions-a-thing-of-the-past/article1759571/).

23. Department of Justice (Canada), The Anti-terrorism Act (accessed 30 June 2011 at http://justice.gc.ca/antiter/sheetfiche/impact-impacte-eng.asp).

24. Government of Canada, *Securing an Open Society: Canada's National Security Policy*, Ottawa: Privy Council Office, April 2004 (accessed 30 June 2011 at http://www.pco-bcp.gc.ca/docs/information/publications/natsec-secnat/natsec-secnat-eng.pdf).

25. Government of Canada, Canada's International Policy Statement: *A Role of Pride and Influence in the World: Overview* (accessed 25 May 2011 at http://merln.ndu.edu/whitepapers/Canada_2005.pdf).

26. Government of Canada, *A Role of Pride and Influence in the World: Defence* (accessed 25 May 2011 at http://www.forces.gc.ca/admpol/downloads/Canada_Defence_2005.pdf).

27. Government of Canada, Canada First Defence Strategy (accessed 25 May 2011 at http://www.forces.gc.ca/site/pri/first-premier/index-eng.asp).

28. Government of Canada, *Canada's Northern Strategy: Our North, Our Heritage, Our Future*, (accessed 30 May 2011 at http://www.canadainternational.gc.ca/eu-ue/policies-politiques/arctic-arctique.aspx).

29. Member states include Canada, the United States, Denmark, Iceland, Norway, Sweden, Finland, and Russia.

30. Bob Weber, "Arctic Treaty to Push Canada to Upgrade Search and Rescue: Experts," *The Globe and Mail*, 4 January 2011, (accessed 5 January 2011 at http://www.theglobeandmail.com/news/politics/arctic-treaty-to-push-canada-to-upgrade-search-and-rescue-experts/article1857858/). University of British Columbia professor Michael Byers interprets the Council members' bid for exclusivity as "essentially saying, 'No, we have matters under control. We are making laws for this area.'"

31. Canadian Security Intelligence Service, *2009–2010 Public Report* (accessed 13 July 2011 at http://www.csis-scrs.gc.ca/pblctns/nnlrprt/2009–2010/rprt2009–2010–eng.asp#tblfn).

32. Quoted in Carl Meyer, "National Security Cabinet Now the Top Foreign Affairs Issue in Cabinet," *Embassy*, 25 May 2011 (accessed 26 May 2011 at http://www.embassymag.ca/page/view/cabinet-05–25–2011).

33. NATO Public Diplomacy Division, Financial and Economic Data Relating to NATO Defence, 10 March 2011 (accessed 20 June 2011 at http://www.nato.int/nato_static/assets/pdf/pdf_2011_03/20110309_PR_CP_2011_027.pdf).

34. Croci, p. 137.

35. National Defence, Backgrounder: Recruiting and retention in the Canadian Forces, 27 May 2011 (accessed 28 May 2011 at http://www.forces.gc.ca/site/news-nouvelles/news-nouvelles-eng.asp?id=3792).

36. The MSOCs were mandated by the 2004 national security strategy Securing an Open Society, and bring together representatives from DND/CF, the RCMP, Transport Canada, CBSA, and the Department of Fisheries and Oceans/Coast Guard.

37. Dave Simms, 'Why Canada Should Care about a U.S. Debt Default," CBC News, 12 July 2011 (accessed 12 July 2011 at http://www.cbc.ca/news/business/story/2011/07/12/f-us-default-canadian-effects.html).

38. Office of the Prime Minister, "PM and U.S. President Obama Announce Shared Vision for Perimeter Security and Economic Competitiveness between Canada and the United States," 4 February 2011 (accessed 30 May 2011 at http://pm.gc.ca/eng/media.asp?id=3931).

39. NATO, "A 'Comprehensive Approach' to Crisis Management" (accessed 2 June 2011 at http://www.nato.int/cps/en/natolive/topics_51633.htm?selectedLocale=en).

40. Government of Canada, *Canada's Cyber Security Strategy: For a Stronger and More Prosperous Canada*, Ottawa, 2010 (accessed 16 June 2011 at http://www.publicsafety.gc.ca/prg/ns/cbr/_fl/ccss-scc-eng.pdf).

41. Postmedia News, "McKay Meets with Rangers Ahead of Arctic Exercises," 24 July 2011 (accessed 25 July 2011 at http://www.globalnews.ca/world/SOMNIA/5152110/story.html).

42. Tonda MacCharles, "Why Michaëlle Jean's Heart Belongs to Haiti," *Toronto Star*, 16 January 2010 (accessed 20 May 2010 at http://www.thestar.com/news/world/article/751555--why-michaelle-jean-s-heart-belong-to-haiti).

43. Campbell Clark, "Solid Americas Strategy Would Reap Big Rewards for Canada," *The Globe and Mail*, 24 May 2011 (accessed 25 May 2011 at http://www.theglobeandmail.com/news/national/time-to-lead/solid-americas-strategy-would-reap-big-rewards-for-canada/article2032365/).

44. Royal Danish Air Force (Tactical Command), Press Release, 2 July 2009. See Canadian-American Strategic Review, *Denmark's Arctic Assets and Canada's Response*, July 2009 (accessed 2 June 2011 at http://www.casr.ca/id-arcticviking-challenger.htm).

45. Gregor Peter Schmitz, "Germany Can Do More," *Der Spiegel*, 26 March 2008 (accessed 18 May 2011 at http://www.spiegel.de/international/world/0,1518,543480,00.html).

46. There are indications that Canada may terminate its participation in the NATO airborne early warning capability, ostensibly on cost grounds. See James Cudmore, "Canada to Pull out of Key NATO Air Defence Program," CBC News, 9 June 2011 (accessed 20 June 2011 at http://www.cbc.ca/news/politics/story/2011/06/09/pol-nato-awacs-canada-mackay.html).

47. Michelle Collins, "Foreign Affairs Hit with $639 Million in Cuts," *Embassy*, 18 March 2011 (accessed 22 July 2011 at http://www.embassymag.ca/page/view/foreign_affairs_cuts-3-18-2009).

48. Campbell Clark, "Harper's Tough Talk on the Arctic Less Stern in Private," *The Globe and Mail*, 12 May 2011 (accessed 14 May 2011 at http://www.theglobeandmail.com/news/politics/harpers-tough-talk-on-the-arctic-less-stern-in-private/article2020615/).

49. Clark (12 May 2011).

50. Official explanations for the severing of diplomatic ties included fears for the safety of Canadian diplomats in Iran, as well as Iran's support for terrorism, its implacable hostility toward Israel, and it domestics human rights record. See "Canada Severs Diplomatic Ties with Ran, Citing Safety Concerns," CTV News, 7 September 2012 (accessed 10 October 2012 at http://www.ctvnews.ca/canada/canada-severs-diplomatic-ties-with-iran-citing-safety-concerns-1.946127).

51. See Kenneth Whyte, "How He Sees Canada's Role in the World and Where He Wants to Take the Country," *Maclean's Magazine*, 11 July 2011, p. 18.

52. Quoted in Collins (18 March 2011).

53. Jordan Michael Smith, "Reinventing Canada: Stephen Harper's Conservative Revolution," *World Affairs*, March/April 2012, pp. 23–24.

54. In May 2011, Canada signed an agreement on the coordination of search-and-rescue efforts with other members of the Arctic Council, thus signaling the internationalization of what had previously been a national security obligation.

55. Brian Stewart, "The New 'Harper Doctrine' in Foreign Relations," CBC News, 16 June 2011 (accessed 18 June 2011 at http://www.cbc.ca/news/canada/story/2011/06/16/f-vp-stewart.html).

56. Don Macnamara and Ann Fitz-Gerald, "What We Learned in Afghanistan," *Ottawa Citizen*, 25 July 2011 (accessed 25 July 2011 at http://www.ottawacitizen.com/news/canada-in-afghanistan/What+learned+Afghanistan/5153028/story.html).

8 Japan's National Security Discourse: Post–Cold War Paradigmatic Shift?

Chris Hughes

Introduction: From Unique to Convergent Discourses of National Security?

Japan in the post–World War 2 period has long been thought of, and indeed openly thought of itself, as an outlier amongst the developed powers in its national security problématique, discourses, and practices. Japan's being bound by Article 9 of its so-called 1946 peace constitution has meant that its outlook on national security has been distinguished by a strong questioning of the efficacy of and reluctance to utilize military power.[1] In fact, to this day the Japan Self Defense Forces (JSDF). established in 1954, have yet to directly inflict a single lethal casualty on any adversary (despite attaining the status of one of the most capable militaries in East Asia. if not globally).

Japan's outlier status has been explained by a variety of viewpoints in academic and policy discourse. For some analysts and policy-makers, Japan has stood as a "structural anomaly," which has been able to preserve its relatively demilitarized posture in the use of force simply because it has been able to rely on U.S. security guarantees under the bilateral security treaty, and thus in effect buck pass to its ally on the real heavy-lifting for military security issues.[2] For others, Japan's stance is explained by attachment to deeply ingrained pacifist or antimilitaristic norms, generating a nearly unique national security culture highly resistant to the use of military force.[3]

In turn, Japan's debate on national security, both inside and outside Japan itself, has only intensified in recent years, with signs of the beginnings of a significant shift in its demilitarized stance. As Japan's external—and to some extent internal—security challenges have mounted in the post–Cold War period, this has led to a questioning of the traditional postwar path of Japanese security policy in terms of the functioning of the U.S.–Japan alliance and the JSDF's own missions and capabilities. The JSDF in East Asia has found itself increasingly upgrading its qualitative capabilities in response to North Korean nuclear and missile provocations, and the far greater looming challenge of China's military modernization. Moreover, globally, the JSDF post–9/11 until 2009 found itself dispatched for the first time in the postwar period to support the United States and international community in on-going conflicts in Afghanistan and Iraq; and since 2008 it has been dispatched as far away as the Gulf of Aden on international antipiracy operations, accompanied by Japan's building of its first overseas military base since the wartime period.[4]

Japan's comparatively recent military proactivity has been driven by and further fuelled transforming discourses on national security, and initiated questions over whether its security trajectory is now starting to finally converge with those of other developed states and becoming what is often termed in the Japanese discourse a "normal" military power.[5] For those traditionally viewing Japan as set to shift in this direction in line with structural pressures, Japan's recent military ambitions reflect the start of inevitable, and often welcome, convergence with other developed industrial and military powers. However, for those traditionally ascribing to the view that antimilitarism and caution about external military activities is ingrained in Japanese society and consequently policy-making, Japan's recent scaling back since the Iraq deployment of out-of-area military commitments is evidence of continued caution and adherence to past trajectories.[6]

The objective of this chapter is to explore these Japanese past and newly emergent discourses and policies in more depth, along with interpretations of the recent changes in its military doctrine and alliance cooperation with the United States, and to consider how these inform us of the degree to which Japan may or may not be converging on a more "normal" security trajectory. Japan's security debate matters because the decisions taken on how to utilize its not inconsiderable military power arguably hold the key to the United States' ability to maintain military hegemony and the overall power balance in the Asia-Pacific, and not least in reacting to the rise of China's military power. In particular, Japan's willingness to cooperate with and facilitate U.S. power in the region will be crucial for President Obama's plans to "pivot" or "rebalance" toward Asia.

Japan's View of Its Fundamental Security Predicament: From Edo to Shōwa

Japan has experienced historically the same set of perceived national vulnerabilities and concomitant national interests and security objectives from the period of the Edo Bakufu (or Tokugawa Shogunate [1603–1868]) through to the Meiji (1868–1912), Taishō (1912–1926), and early Shōwa (1926–1989) periods. Japan's elites have traditionally perceived their polity to be one that is inherently vulnerable due to limited natural and energy resources, land space and population, and to be positioned at the juncture of a uniquely difficult set of regional and global environments. Japan's rulers at the start of the Edo period, fearful of external regional instability on the East Asian continent and of the encroachment of Western imperialism, chose the path of autarky and imposed a policy of near total self-isolation (*sakoku*) from the outside world. Japan's closure endured for close to two hundred years, until the country was opened forcibly in the mid-19th century by the arrival of the Western powers.

Japan's political elites, after a degree of internal struggle, rapidly concluded that autarky was no longer a viable option, as it left Japan backward in economic and military development and as a disadvantaged late-starter power in a rapidly deteriorating regional environment. Japan's leaders feared that their country might follow the path of China and become gradually dismembered by and dependent upon the Western imperial powers. Japan in response to these challenges forged a new grand strategy, seeking to convert itself into a modern nation-state and imperial power capable of resisting external incursions. Japanese leaders were determined to catch-up with the early-starter Western powers and subsequently achieved rapid industrialization and the building up of naval and land forces powerful enough to gain victory in the Sino-Japanese War of 1894–1895 and in the Russo-Japanese War of 1904–1905; to secure for itself colonies in Taiwan, Korea, and China; and from 1902 onwards to forge an alliance relationship with Great Britain as the hegemonic power of the day.[7]

Japan's grand strategy, however, came unstuck in the early 20th century as the international system was perceived to turn once again to its disadvantage. Despite Japan's participation on the victorious Allied side in World War 1, it received unfavorable treatment at the Paris Peace Conferences (1919), within the League of Nations, and at the Washington Naval Treaty (1922). Japan's attempts at further colonial expansion in Manchuria—a perceived necessity from the Japanese perspective to halt instability in China and Soviet aggression—were censured by the Western imperial powers. Japan's leaders, viewing the international institutions as biased against their attempts to protect their national autonomy through "defensive" imperial expansion regionally, and perceiving a shift in the global balance of power to the revisionist Axis Powers, concluded a Tripartite Pact with Nazi Germany and Italy in 1940. Japan subsequently sought to dislodge the Western imperial powers from the East Asia region in order to secure an exclusive economic and resource space for its national security—a new form of autarky. Japan's strategic choice was, of course, to prove disastrous. Stunning initial Japanese victories and the establishment of the Greater East Asian Co-Prosperity Sphere in 1941–1942 were to be followed by conventional military defeat and the atomic bombings of Japan at the hands of Allied Powers by 1945.[8]

Japan's total defeat, subsequent economic devastation, loss of independence under the U.S.-dominated Allied Occupation (1945–1951), the process of demilitarization embodied in the acceptance of Article 9 of the Constitution, alienation of its Asian neighbors, and the oncoming Cold War reestablished all of its old vulnerabilities and forced its leaders to search for a new grand strategy in the postwar period. Japan's policy-makers rehearsed a number of new and past strategic discourses. On the left of the political spectrum, the Japan Socialist Party (JSP) and Japan Communist Party (JCP) charged that the domestic failings of prewar democracy and resultant ultra-nationalism had been responsible for leading Japan into the calamity of war, and that Japan could only restore its independence by following a new path of the completion of the process of demilitarization and democratization, the pursuit of unarmed neutrality, and restoring ties with former colonies in East Asia. At the other extreme, conservative nationalists, or so-called Revisionists, argued that Japan should restore its autonomy through reprising a prewar strategy of building up national military capabilities, creating shifting and multiple alliances, revising Article 9 and its constraining influence on the exercise of military power, and rejecting foreign troops based on Japanese soil.[9]

In the final eventuality, these contending discourses were reconciled, or at the very least marginalized, by the forging of a compromise position led by Japan's conservative Pragmatists under the direction of Prime Minister Yoshida Shigeru (1946–1947, 1948–1952). Yoshida was committed to restoring Japan's position amongst the ranks of the great powers, but his strategy rejected increasing military spending and large-scale rearmament as unfeasible given the perilous state of the Japanese economy and the anticipated domestic political opposition from the JSP, JCP, the wider public, and East Asian states. Yoshida and the Pragmatists perceived, instead, that the reconstruction of the civilian economy and technological prowess were to be the future prerequisites for ensuring national autonomy. Japan was to reemerge as a merchant state, not a samurai one; and national wealth was to be rebuilt through maritime trade and the gradual regaining of markets in the United States, Europe, and East Asia. The Pragmatists were highly realistic and nationalist in outlook and did not reject altogether the role of military power in ensuring national autonomy, and they were prepared to contemplate more significant rearmament and Japan's reemergence as an autonomous military power in the future.

In order implement this new grand strategy, later termed Yoshida Doctrine, Prime Minister Yoshida chose to reprise another former Japanese strategy from the early 20th century of moving closer to the established hegemonic power of the day. Japan was thus committed under Yoshida to alignment—although not necessarily alliance—with its former adversary, the United States, seeking and signing the 1951 Security Treaty. The bilateral security treaty initiated an implicit grand strategic bargain. Japan was obliged to provide the United States with bases to enable the projection of U.S. power onto the East Asian continent to pursue Cold War containment.[10] In separate agreements, Japan committed itself to assume a degree of responsibility for national self-defense through light rearmament and the formation of the National Police Reserve in 1950 and the National Safety Force in 1952—the forerunners of the JSDF. In return, Japan gained effective (if not explicit, until the revised security treaty of 1960) guarantees of superpower military protection, including forward-deployed forces in Japan and the extended U.S. nuclear "umbrella" deterrent.

In accepting these security arrangements, Japan further gained U.S. assent for the ending of the occupation and thus restoration of its independence (although the United States retained administrative control of Okinawa until 1972). Japan's postwar alignment also brought it (as originally calculated by Yoshida, based on his view that the United States needed Japan as a Cold War bastion of capitalism in East Asia) economic security guarantees in the form of special economic dispensations, including access to the U.S. market, economic aid and international economic institutions, and technology transfers. Hence, through U.S. sponsorship, Japan was able regain its place in the international community and, furnished as it was with U.S. military protection, was free to pursue its goal of economic reconstruction. In addition to meeting the challenges of its postwar international vulnerabilities, Yoshida's decision to entrust, in large part, U.S. military security enabled the management of domestic controversies over Japan's future military stance: the JSP and JCP objecting to the U.S.–Japan security treaty as abetting U.S. imperialism but robbed of political leverage by the avoidance of large-scale rearmament; and the Pragmatists and Revisionists eventually joining forces through the formation in 1955 of the long-governing Liberal Democratic Party (LDP) (holding power for a near uninterrupted period of more than fifty years until 2009) in order to beat off the challenge from their common left-wing adversaries, and then reaching a consensus on the necessity of maintaining the U.S.–Japan security treaty.

Japan's LDP was subsequently to sit relatively comfortably under the economic and security benefits of U.S. hegemony for most of the Cold War period. Japanese policy-makers were aware of potential costs associated with this alliance and that Japan's commitment should not be unconditional. Yoshida in particular hoped U.S. alignment would be a temporary expedient, stating famously that "the day [for rearmament] will come naturally when our livelihood recovers. It may sound devious (*zurui*), but let the Americans handle [our security] until then."[11] The principal costs for Japan of alignment were the alliance dilemmas of abandonment and most especially entrapment by the United States in its regional and global security strategy. During the Cold War, Japanese policy-makers feared abandonment the least, because they perceived that the United States valued Japan too highly as a central component of its containment strategy in East Asia. However, for successive Japanese administrations, U.S. military dependence, and the knowledge that the U.S. interests as a global power could supersede those regional interests related to Japan, has engendered residual fears that the United States' commitment to defend Japan might eventually wane, exposing its vulnerabilities. The major concern for Japanese policy-makers, though, has been that of entrapment, either through requests for Japan to provide reciprocal defense for U.S. territory or to engage in U.S.-led bilateral or collective security arrangements in the Asia-Pacific and beyond.

Consequently, Japanese policy-makers ever since the signing of the security treaty have tempered the risks of entrapment through hedging tactics vis-à-vis growing U.S. demands for security burden-sharing. Japanese policy-makers have stressed to their U.S. counterparts that Article 9 has been interpreted to enable Japan solely to exercise the right of individual national self-defense, and not the exercise of the right of collective self-defense, and thus that Japan cannot exercise armed force outside its own territory to defend its U.S. security treaty partner or assist other U.S. allies or other states. Japan also chose to emphasize U.S. military cooperation under the security treaty in line with Article 5 (the immediate defense of Japan) rather than Article 6 (the maintenance of international peace and security in the Far East). Japan refused to develop the type of military capabilities that might mean it could be called upon to participate in U.S.-led expeditions. Japan in this way avoided pressure to join other U.S. allies, such as South Korea, in participating in the Vietnam War.

Moreover, Japan, even in the latter stages of the Cold War, as it embarked on a qualitative and quantitative build-up of JSDF capabilities—the Japan Ground Self-Defense Force (JGSDF) shifting its heavy forces and tank deployments to Hokkaidō; the Japan Air Self-Defense Force (JASDF) acquiring E-2C early-warning aircraft and F-15 advanced interceptors; and the Japan Maritime Self-Defense Force's (JMSDF) large numbers of destroyers and P-3Cs for antisubmarine warfare—that all served to counter the USSR's ability to threaten the airspace and sea lanes around Japan and thus supported overall U.S. strategy in Northeast Asia, carefully avoided the integration of JSDF capabilities and missions with those of the U.S. military for fear of entrapment.[12] Japan and the United States did embark on the first steps toward the direct coordination of their respective military roles through the formulation of the 1978 Guidelines for Japan-U.S. Defense Cooperation, including under Article 5 of the security treaty tactical planning, joint exercises, and logistical support, and under Article 6 cooperation in regional contingencies and Japan's patrolling of the Sea Lines of Communication (SLOC). However, in practice, Japan concentrated only on research into Article 5–type contingencies for its own immediate defense.

Japan's reserved alliance stance was reinforced by a range of antimilitaristic principles or taboos, nonbinding in legislation but normatively ingrained in society at large. These included the Three Non-Nuclear Principles of 1967 (not to possess, produce, or introduce nuclear weapons); the total bans on the export of military technology introduced in 1967 and 1976; the insistence on peaceful use of space in 1969; and the 1 percent of GNP limit on defense expenditure introduced in 1976. In turn, these constraints on the exercise of Japanese military power were reinforced by the fact that the LDP, despite holding power continually and being able to nudge forward Japan's military role alongside the United States, had to contend with stiff opposition from the JSP as the principal opposition party.

Hence, even though Japan's then Prime Minister Suzuki Zenkō was able by 1981 to refer for the first time publicly to U.S. security arrangements as an "alliance," Japan's actual military security role was one that had been geographically limited to the area immediately surrounding Japan and functionally limited to providing a defensive "shield" to support the U.S. offensive "spear" in Northeast Asia; and if Japan made any contribution to wider regional security it was an indirect one through the mechanism of the bilateral alliance and support for the U.S. presence in East Asia.

Japan's "Grand Strategy" under Stress in the Heisei Era: International Systemic Challenges and Renewed Security Discourses

The pursuit of the Yoshida Doctrine and the maintenance of a highly asymmetrical U.S. alliance worked largely to Japan's benefit during the Cold War. However, in the post–Cold War period, or Heisei era (1989–), there have been indications that Japan's grand strategy is once again becoming untenable in the face of regional and global challenges.

Japan's East Asian security environment has significantly deteriorated with the ending of the certainties of bipolarity during the Cold War and the more fluid environment introduced by fleeting U.S. unipolarity, and then the transition to a possibly more multipolar international system. The most immediate challenge Japan has faced is the miscreant state of North Korea. Japanese anxieties had focused on North Korea's development of nuclear weapons alongside the strengthening of its ballistic missile capabilities to produce a credible *force de frappe*. Japan has seen North Korea progressively push forward its nuclear capabilities through a series of nuclear crises since the mid-1990s, culminating in its conduct of nuclear tests in October 2006 and May 2009. In turn, Japan's vulnerability to missile attack was demonstrated by the Taepdong-1 shock of August 1998, when North Korea test-fired a missile over Japanese airspace; by North Korea's conducting of a series of missile tests in the Sea of Japan in July 2006; and then again by North Korea launching a Taepdong-2 missile test over Japan in April 2009. North Korea then conducted further rocket launches in April and December 2012. Added to this, Japanese defense officials have been concerned about the incursion of North Korean "spy ships" (*fushinsen*) on espionage missions and the risks of guerrilla attacks upon key facilities such as nuclear power installations on the Sea of Japan coastline.[13]

However, it is China that poses the greatest challenge for Japanese security over the medium to long terms. Japan has been concerned at China's modernization of its conventional and nuclear forces since the early 1990s. Japanese concerns focus, though, not just on China's military modernization per se but upon signs that it is now willing to project military power beyond its borders in support of its national interests. Japan is aware that China could disrupt its SLOCs with only a small blue-water surface, submarine, and amphibious naval capacity and through the assertion of its territorial claims to the South China Sea. China's constant dispatch of research ships and warships into Japan's Exclusive Economic Zone (EEZ) around the disputed Senkaku (Diaoyutai) Islands in the East China Sea are taken as evidence of aggressive intent.[14] Japan's concerns about China have been compounded by the latter's natural gas exploration activities in the East China Sea, in fields abutting onto Japan's Exclusive Economic Zone, and the inability by both sides to forge a working pact on joint development of the fields. Japan–China maritime tensions escalated in 2010 with Japan highlighting the passage of a large People's Liberation Army Navy (PLA-N) flotilla (or armada, as described in the Japan media) passing close to Japan's southernmost islands. Japan and China subsequently clashed over the Senkakus in late 2010. The Japan Coast Guard (JCG) arrested the captain and crew of a Chinese trawler that attempted in September to ram its vessels in the disputed area, leading to China reacting by curtailing diplomatic ties with Japan and temporarily halting exports of strategic rare earths materials. Japan-China relations then further deteriorated over the Senkakus from 2012 onwards, with the Japanese government's nationalization of the islands through purchasing two of them from their private owner in order to prevent their potential falling into the hands of domestic right-wing political forces, but which then triggered a ramping up of Chinese military activity around the islands.

Japan–China security relations have been further complicated since the mid-1990s by the issue of Taiwan and Sino–U.S. strategic competition. Japan viewed the 1995–1996 Taiwan Straits crisis and China's intimidation of Taiwan through the test firing of ballistic missiles with alarm, seeing it as another indication of China's willingness to project military power in pursuit of its national interests; to possibly challenge the United States militarily in the region over the longer term; and even to use ballistic missiles to strike against U.S. bases in Japan and against rear area support facilities provided by Japan in the event of a full-blown conflict resulting from any Taiwanese move to declare independence. Japanese security planners apparently fear as well that in a Taiwan Straits crisis China might attempt to seize offshore islands such as parts of Okinawa Prefecture in order to disrupt U.S.–Japan military cooperation.

Japan in the post–Cold War period has now come to face for the first time involvement in a series of global security challenges and responsibilities, especially with the realization that U.S. hegemony and willingness to provide security dispensations are no longer boundless. Japan's response to the 1990–1991 Gulf War, which took the form of underwriting the war financially by providing US$13 billion, rather than a human contribution in the form of JSDF dispatch, was the subject of U.S. and international criticism, given the scale to which Japan also relied on stability in the Middle East for its economic welfare. It also made Japanese policy-makers aware of the need for a more proactive stance in supporting international efforts to respond to major post–Cold War security crises.[15] Similarly, Japan, in the wake of 9/11, was made aware of the threat of transnational terrorism and the demands from its U.S. ally to support the United States and the international community in efforts to expunge this threat. Japanese policy-makers have also been in accord with their U.S. counterparts on the need to halt the proliferation of WMD to other states or even possibly terrorist groups.

Japan's preferred role in responding to this post–Cold War and post-9/11 security agenda has clearly been a nonmilitary one, which relies on economic power and diplomacy. Japan has persevered with diplomacy toward North Korea and engagement toward China. Japan has taken similar lessons from the war on terrorism and the Iraq war. As will be seen in the next section, Japan sought roles in supporting the Afghan conflict and in Iraq that emphasized the use of economic power, postconflict reconstruction and state-building. Moreover, its policy-makers maintain strong reservations concerning the utility of military power in bringing about a resolution to the multicausal phenomenon of terrorism in Afghanistan, and the ability of the United States to reconstruct and stabilize postwar Iraq.[16] Nevertheless, Japanese policy-makers are increasingly aware of the limits of traditional low-profile diplomatic and economic responses to security issues and realize that in the final calculation,

Japan may now need to line up more closely with U.S. efforts to deter militarily North Korea and China and to deal with other challenges globally.

In turn, Japan's shifting external security environment has triggered a shifting set of domestic political conditions and security discourses that challenge the maintenance of the former status quo in security policy. The end of the Cold War has challenged the legitimacy of the existing political forces in Japan. The LDP since the early 1990s has overseen a decline in Japan's economic fortunes and further questioning of its legitimacy to govern as its anti-communist credentials have become less relevant. Similarly, the SDPJ's antimilitaristic stance has become less credible as Japan has been faced by a series of extant security crises. The consequence has been that the SDPJ has largely imploded as a political force over the last decade, while the LDP has searched for new sources of domestic legitimacy in its economic program and external security policy. The LDP was to eventually fall from power, largely due to economic issues, and gave way as the governing party in September 2009 to the Democratic Party of Japan (DPJ), which had emerged as the main opposition party over the previous decade. The LDP was then to regain power in December 2013. This process of the fall then rise of the LDP in power has been accompanied by a consequent new fluidity in Japanese thinking about security policy.

The extremes of the Japanese security debate have partially reemerged on either side of the spectrum: Japanese right-wing nationalists call for significant remilitarization, including the acquisition of nuclear weapons, to guarantee Japan's future independence; whereas the remnants of the far left-wing continue to call for a neutral and demilitarized stance. However, the Japanese mainstream has tended to converge on three contending yet in many ways overlapping views of the appropriate security trajectory. The Pragmatists in the LDP continued to hold overall sway until 2009 and clung to a modified version of the Yoshida Doctrine of a Japanese military low-profile pushed incrementally forward in step with expanding U.S.–Japan regional alliance functions. The Revisionists increasingly reasserted themselves, however, under the likes of Prime Ministers Kozumi Junichirō (2001–2006), Abe Shinzō (2006–2007; 2012–), and Asō Tarō (2008–2009), arguing for a more assertive Japanese military role regionally and globally, although again through the mechanism of the U.S.–Japan security treaty and arguing for an alliance somewhat akin to the UK–U.S. security partnership. The DPJ for its brief period in power argued that Japan should certainly take a greater international security role and strengthen the U.S. alliance. But the DPJ has also cautioned that Japan should exercise more of its military capabilities via multilateral institutions such as the UN, expand multilateral security cooperation with East Asian neighbors, and take care not to overly stretch U.S. military cooperation outside the East Asian region. Japan has thus not yet reached a post–Cold War consensus in its security discourse but is clearly edging toward a new acceptance of the need to play a greater military role, and that much of this role should be exercised through a revitalized U.S.–Japan alliance.

Japan's Shifting National Security Policy: Strengthened Alliance and JSDF Capabilities

Japan's first wave of security policy transformation began in the mid-1990s and was focused on the U.S.–Japan alliance's regional security functions. Japan's recognition of the inadequacy of the alliance and its role within it as a mechanism to respond to regional security challenges was initiated by the North Korean nuclear crisis of 1994–1995. Japan at this time faced U.S. requests for active support in the event of a conflict on the Korean Peninsula, including rear area logistical support. Japanese policy-makers were unable to respond effectively, though, due to their previous reluctance to consider Article 6–type cooperation for regional contingencies under the bilateral security treaty. The result was to reveal the hollowness and lack of military operability of the so-called alliance; to induce a full-blown crisis of political confidence bilateral alliance relationships; and to raise the prospect of Japan's abandonment by the United States as an unreliable ally.[17] Japan–U.S. security tensions in this period were also compounded by tensions over U.S. bases in Okinawa prefecture, and most especially crimes perpetrated by U.S. servicemen and the environmental impact of the bases.

Japanese policy-makers' fear that the postwar strategic bargain with the United States risked becoming undone meant that they took initiatives to revise Japan's military doctrine and to redefine the alliance. Japan in November 1995 issued a revised National Defense Program Outline (NDPO)—the document that sets out Japan's military doctrine alongside the necessary force structure—which was significant in stressing the need for stronger U.S.–Japan alliance cooperation, and inserted a new clause to state that if a situation should arise in areas surrounding Japan (*shūhen*) that impacts upon national peace and security, then Japan should seek to deal with this in cooperation with the UN and via U.S.–Japan security arrangements. Japan and the United States then issued a Joint Declaration on Security in April 1996 that opened the way for a revision between 1996 and 1997 of the original 1978 Japan–U.S. Guidelines for Defense Cooperation. The revised guidelines specified for the first time the extent of Japanese logistical support for the United States in the event of a regional contingency (*shūhen jitai*), and thereby switched the agreed emphasis of the alliance cooperation from Article 5 to Article 6 of the security treaty.[18]

Japan post-9/11 then moved to a new global security agenda. Japan passed through the National Diet in October 2001 an Anti-Terrorism Special Measures Law (ATSML) that enabled the dispatch from November 2001 until January 2010 of JSDF units to the Indian Ocean area to provide logistical support to U.S. and multinational coalition forces engaged in Afghanistan, and in particular MSDF refueling of U.S. and coalition ships in the Indian Ocean.[19] Japan then passed in July 2003 an Iraq reconstruction law that enabled the dispatch of JSDF from 2004 to 2006 on noncombat reconstruction activities in the southern city of Samawah. The GSDF worked under the protection of UK, Dutch, and Australian forces. Japan predicated these laws and JSDF out-of-area dispatch upon linkages to relevant UN resolutions, and thus these security activities are strictly outside the geographical and functional scope of the U.S.–Japan security. Nevertheless, it is clear the principal impetus behind the ATSML and Iraqi reconstruction law was Japanese attempts to strengthen the range of U.S.–Japan alliance activities outside East Asia.

Japan's focus on an expanded regional and global security role since the early 1990s has not been exclusively on the U.S.–Japan alliance. Japan's failure to respond to U.S. and international expectations for JSDF dispatch during the Gulf War of 1990–1991 provided the opportunity for the passing in June 1992 of the International Peace Cooperation Law (IPCL) to allow JSDF dispatch on noncombat reconstruction UN PKO. Japan has taken part in UN PKO in Cambodia (1992–1993), Mozambique (1993–1995), Rwanda (1994), the Golan Heights (1996–2013), East Timor (2002–2004), Haiti (2010), and South Sudan (2011–) with other small deployments to support UN missions in Nepal and Sudan. Japan in 2002 also unfroze provisions in the ICLP that allow JSDF participation in core UN PKO, including the monitoring of ceasefires, collection of weapons, and exchange of prisoners. Nevertheless, Japan's main preoccupation has not been with UNPKO in recent years, with for instance less than 30 of its 240,000 personnel deployed on these missions in 2009.[20]

Japan in deploying the JSDF overseas has thus begun to shift away from the rigid tenets of the Yoshida Doctrine, but it is notable that Japan has retained the Pragmatists' line by continuing to hedge carefully in its U.S. commitments. Japan in the revised Guidelines for Defense Cooperation has stressed that their activation is predicated upon the concept of situational need rather than strict geographical demarcations, thus introducing an element of strategic ambiguity as to whether the scope of the revised guidelines necessarily covers Taiwan. Japan in responding to the war on terrorism enacted individual laws to enable JSDF dispatch to support U.S.-led coalitions in Afghanistan and Iraq and predicated these laws on UN resolutions, thereby creating constitutional and legal firewalls between these operations and the range of support that it is prepared to offer the United States under the bilateral security treaty. Moreover, Japan has restricted the expansion of JSDF activities regionally and globally to noncombat missions and as yet has left intact its constitutional prohibition on the exercise of collective self-defense.[21] Furthermore, the DPJ after it took power was far more wary of committing the JSDF to new out-of-area operations, withdrawing the JMSDF from the Indian Ocean, although maintaining a deployment of MSDF destroyers and P-3Cs in the Gulf of Aden as part antipiracy missions and constructing in 2011 a permanent base for these in Djibouti.

Nevertheless, Japan's military planners still remain committed to steadily pushing forward long-term plans for a more activist global alliance with the United States. Japan and the United States in 2006 concluded a Defense Policy Review Initiative (DPRI), which then resulted in a series of Security Consultative Committee statements, the main mechanism for bilateral security planning, which were significant in stressing that the United States and Japan now shared not just common regional but also now *global* strategic objectives. The DPRI included plans for the relocation of the command functions of U.S. Army I Corps, a rapid-reaction force typically operating from the Asia-Pacific to the Middle East, from Washington State to Camp Zama near Tokyo. The ramification of this were that Japan would serve as a frontline command post for U.S. global power projection to as far away as the Middle East, thus marking an implicit breaching of the interpretations of the scope of the U.S.–Japan security treaty and U.S. bases as covering only Japan and the Far East. Japan has complemented this realignment by moving its own newly established JGSDF Central Readiness Force, with rapid-reaction capabilities, to Camp Zama. In addition, Japan agreed to the establishment of a Bilateral Joint Operations Coordination Centre (BJOCC) at Yokota air base to collocate the ASDF's command and control functions of Japan's emergent Ballistic Missile Defense system with those of the U.S. military in Japan, and for the United States to deploy additional and complementary BMD assets around Japan, including an X-band radar system at Kashiri in Aomori Prefecture and *Patriot* Advanced Capability (PAC)-3 systems. Hence, Japan has now largely abandoned its former Cold War fear of integration of the JSDF with the U.S. military.[22]

In the meantime, Japan has worked on upgrading its own national defense doctrines and JSDF capabilities. The MOD in December 2004 released its revised National Defense Program Guideline (NDPG), to replace the 1996 NDPO, and it stressed the need for Japan to concentrate its security activities on the area spreading from the Middle East to East Asia. Japan's policy-makers have subsequently worked to change the nature of the JSDF as essentially a force designed to ward off the USSR during the Cold War and instead convert it into a lighter, more mobile force to respond to regional and global contingencies, and more interoperable with the U.S. and other militaries. Consequently, the 2004 NDPG reduced the number of GSDF tanks, ASDF interceptors, and JMSDF destroyers and emphasized the need for more qualitatively advanced weaponry.[23]

The latest iteration of the NDPG in 2010 then went further in reshaping the JSDF, finally abandoning the 1970s concept of the basic defense force in favor of a new dynamic defense force. The basic defense force concept was a hangover from the Cold War period whereby Japan maintained the minimum defense capabilities supposedly just sufficient to defend its territory. In contrast, the dynamic defense force concept is one where the JSDF will calibrate its capabilities more closely, in near balance of power fashion, to counter specific sources of threat and contingencies. In order to counter the emerging maritime threat from China, the MSDF plans to continue to procure DDH light helicopter carriers, to maintain a very substantial fleet of forty-eight destroyers, and an increased submarine fleet of twenty-two vessels. The MSDF will also complete the fitting out of six of its destroyers for a BMD capability. The JASDF will reduce its number of fighters but has compensated for this by a qualitative upgrade through acquiring the F-35 fifth-generation fighter. The JGSDF will continue to cut its number of tanks and artillery but further develop rapid reaction capabilities, with a new emphasis on garrisoning Japan's southernmost islands in the East China Sea—a move again designed to counter possible Chinese incursions.[24] All of these new Japanese capabilities should also assist the U.S.

emerging plans to hedge against China's military rise through the Air-Sea Battle Concept, with Japan providing key BMD, air defense, and intelligence, surveillance, and reconnaissance assets

Japan's defense establishment in this period has not only challenged postwar constraints by acquiring greater power projection capabilities for the JSDF, but also by increasingly eroding a number of key antimilitaristic principles that constrain military planning. Japan has effectively breached the 1969 prohibition on the use of space for military activities by launching since 2003 a series of spy satellites and now declaring that it would use space for defensive rather than nonmilitary purposes per se.[25] Japan's defense production sector has pushed hard in recent years and came close in 2010 to breaching the ban on arms exports in favor of an export license system, desirous as it is of enhanced opportunities for international co-development of new weapons systems with United States and other partners. Japanese policy-makers have turned a blind eye since the 1960s to breaches of the third of the non-nuclear principles by allowing the United States to introduce nuclear weapons into Japan in transit of its navy ships; and more recently a number of high-profile politicians have raised questions about the utility of Japan's acquiring its own nuclear deterrent as U.S. extended nuclear guarantees are feared to possibly slip in face of North Korea's nuclearization. Finally, Japanese policy-makers have come to question the value of constitutional limitations on the use of military force. The LDP in particular, but also elements of the DPJ, have sought to overturn the self-imposed ban on the exercise of the right of collective self-defense because it impedes U.S.–Japan alliance cooperation. The Japanese National Diet has also begun to investigate for the first time the possible formal revision of Article 9 of the Constitution to acknowledge the existence of the JSDF as a military and to specify its responsibilities for national defense.[26]

Reflections: A Radical Turning Point in Japanese Security Policy?

Japan appears to be nearing, if incrementally, a potential paradigmatic shift in the context and trajectory of its postwar security trajectory—moving from a situation of a highly circumscribed defensive posture concerned purely with its own immediate national security to one of a more assertive defense stance, more closely allied with the United States, and extending beyond its own region to include global security functions. In short, the so-called Yoshida

Doctrine appears to be losing its grip on the making of national security policy. Japanese policy-makers and analysts are now considering how far Japan may continue to move along this trajectory.

It is certain that Japan's concerns about the changing international structure and the pressure this exerts on changes for its security policy look unlikely to abate in the short, medium, or long terms. Japan's concerns over North Korea may dissipate if international efforts contain the nuclear issue. However, Japan will still then be confronted more openly by China's seemly inexorable rise, which will continue to drive forward Japanese remilitarization to cope with related territorial and resource threats. At the same time, Japan's ability to shelter behind the United States in the international system will come increasingly into doubt. All of this suggests that Japan's domestic attachment to discourse and norms of antimilitarism will be progressively eroded in the future, reinforcing the cycle of remilitarization.

Japan's exact pathway toward a more remilitarized stance though is less certain. Japan in the past, as made clear in the historical sections of this chapter, has made radical jumps in strategic directions if the international system is perceived to turn against it. For Japan any leap from the current hegemonic power of the United States, however currently weakened, toward band-wagoning with China as the potentially rising dominant power seems unlikely. This is due to the fact that China clearly stands more as a threat at present; that the prospects for U.S. hegemony are really finished are far from clear; and that Japan's policy-makers are inherently risk averse, given past attempts to shift toward revisionist powers. In turn, any attempt by Japan to retreat into isolation appears unlikely, given the historical lessons of this strategy; Japanese strategic dependence on the outside world for economic welfare; and its own lack of national strength to adequately defend itself, short of adopting its own nuclear deterrent.

Hence, the most likely path for Japan to follow is to strengthen its own national capabilities in conjunction with a stronger U.S.–Japan alliance. This will allow Japan to avoid the full costs of an independent defense posture, to maintain a semblance of its post-war antimilitarism, to bolster the weakening U.S. presence in the region, and provide Japan with the assurance to gradually expand outwards its international security role. Japan's progress along this path is unlikely, though, to be smooth. Japan's acceptance of increased alliance responsibilities and security interdependence or dependence on the United States will engender renewed abandonment and entrapment concerns. Japan's policy-makers may at times try to retreat back into essentially a low-profile stance, or defensive realism, to avoid the costs of alliance and simply concentrate on homeland security.[27] At other times, Japan may be obliged to assume more a form of reluctant realism, accepting the need to gradually follow the United States in facing up to regional and global security exigencies.[28] Finally, in other instances, Japan's ambitions as an independent and Great Power may be frustrated by rivalry from China and dependence on the United States generating unpredictability in its defense posture, characterized more by what might be termed resentful realism and a tendency to lash out against its adversaries.[29]

Notes

1. Article 9 of Japan's Constitution states:

> Aspiring sincerely to an international peace based on justice and order, the Japanese people forever renounce war as a sovereign right of the nation and the threat or use of force as means of settling international disputes.
>
> To accomplish the aim of the preceding paragraph, land, sea, and air forces, as well as other war potential, will never be maintained. The right of belligerency of the state will not be recognized.

2. Kenneth N. Waltz, "The Emerging Structure of International Politics," *International Security*, vol.18, no. 2, Fall 1993, pp. 44–79; Jennifer M. Lind, "Pacifism or Passing the Buck? Testing Theories of Japanese Security Policy," *International Security*, vol. 29, no. 1, Summer 2004, pp. 92–121.

3. Thomas C Berger, "From Sword to Chrysanthemum: Japan's Culture of Anti-Militarism," *International Security*, vol. 17, no. 4, Spring 1993, pp. 118–150; P. J. Katzenstein and N. Okawara, "Japan's National Security: Structures, Norms, and Policies," *International Security*, vol. 17, no. 4, Spring 1993, pp. 84–118; Glenn D. Hook, *Demilitarization and Remilitarization in Contemporary Japan*, (London: Routledge, 1996); H. Richard Friman, Peter J. Katzenstein, David Leheny, and Nobuo Okawara, "Immovable Object? Japan's Security Policy in East Asia" in Peter J. Katzenstein and Takashi Shiraishi, eds., *Beyond Japan: The Dynamics of East Asian Regionalism*, (Ithaca, N.Y.: Cornell University Press, 2006), pp. 85–107; Andrew L. Oros, *Normalizing Japan: Politics, Identity and the Evolution of Security Practice*, (Stanford, Calif.: Stanford University Press, 2008); Yasuo Takao, *Is Japan Really Remilitarising?* (Victoria, Australia: Monash University Press, 2008); Y. Izumikawa, "Explaining Japanese Antimilitarism: Normative and Realist Constraints on Japan's Security Policy," *International Security*, vol.35, no.2, Fall 2010, pp.123-160.

4. Christopher W. Hughes, *Japan's Remilitarisation* (London: Routledge, 2009), pp. 35–52, 79–98.

5. Christopher W. Hughes, *Japan's Re-Emergence as a "Normal" Military Power*, (Oxford, Oxford University Press, 2004); Richard J Samuels, *Securing Japan: Tokyo's Grand Strategy and the Future of East Asia* (Ithaca, N.Y.: Cornell University Press, 2007); Kenneth B. Pyle, *Japan Rising: The Resurgence of Japanese Power and Purpose* (New York: Public Affairs, 2007).

6. Paul Midford, *Rethinking Japanese Public Opinion and Security: From Pacifism to Realism?* (Stanford, Calif.: Stanford University Press, 2011).

7. Glenn D. Hook, Julie Gilson, Christopher W. Hughes, and Hugo Dobson, *Japan's International Relations: Politics, Economics and Security*, 3rd ed. (London: Routledge, 2011), pp. 27–31, 182–184.

8. Samuels, *Securing Japan: Tokyo's Grand Strategy and the Future of East Asia*, 13–28.

9. Ibid., 29–37.

10. Hughes, *Japan's Re-Emergence as a "Normal" Military Power*, pp. 21–24.

11. Pyle, *Japan Rising: The Resurgence of Japanese Power and Purpose*, p. 230.

12. Christopher W. Hughes and Akiko Fukushima, "US-Japan Security relations—Toward Bilateralism Plus?," in *Beyond Bilateralism: U.S.-Japan Relations in the New Asia-Pacific*, Ellis S. Krauss and T. J. Pempel, eds (Stanford, Calif.: Stanford University Press, 2003), pp. 61–63.

13. Christopher W. Hughes, "Supersizing the DPRK Threat: Japan's Evolving Military Posture and North Korea," *Asian Survey*, vol. 49, no. 2, March/April 2009, pp. 291–311.

14. Reinhard W. Drifte, *Japan's Security Relations with China since 1989: From Balancing to Bandwagoning?* (London: Routledge, 2002), pp. 33–83; Euan Graham, *Japan's Sealane Security, 1940–2004: A Matter of Life and Death?* (London: Routledge, 2006), pp. 203–220; Richard Bush, *The Perils of Proximity: China-Japan Security Relations* (Washington, D.C.: Brookings Institution Press, 2010).

15. Kent E. Calder, *Pacific Alliance: Reviving U.S.-Japan Relations* (New Haven, Conn.: Yale University Press, 2009), pp. 134–157.

16. Christopher W. Hughes, "Not Quite the 'Great Britain of the Far East': Japan's Security, the U.S.-Japan Alliance and the 'War on Terror' in East Asia," *Cambridge Review of International Affairs*, vol. 20, no. 2, 2007, pp. 325–338.

17. Akihiko Tanaka *Anzen Hosho: Sengo 50-nen no Mosaku* (Tokyo: Yomiuri Shimbunsha, 1997), pp. 323–347; Yoichi Funabashi, *Alliance Adrift*, (New York: Council of Foreign Relations Press, 1999), pp. 280–295.

18. Hughes, *Japan's Re-Emergence as a "Normal" Military Power*, pp. 67–73.

19. Paul Midford, "Japan's Response to Terror: Dispatching the SDF to the Arabian Sea," *Asia Survey*, vol. 43, no. 2, March/April 2003, pp. 331–333.

20. Hughes, *Japan's Remilitarisation*, 84.

21. Hughes, *Japan's Re-Emergence*, 2004, pp. 131–133.

22. Hughes, *Japan's Remilitarisation*, 84

23. Ibid., 40–48.

24. Japan Ministry of Defense, *National Defense Program Guidelines for FY 2011 and Beyond*, pp. 6–7.

25. Pekkanen and Kallender-Umezu. *In Defense of Japan: From Market to the Military in Space Policy.*

26. Hughes, *Japan's Re-Emergence as a "Normal" Military Power*, pp. 102–129.

27. Paul Midford, *Rethinking Japanese Public Opinion and Security: From Pacifism to Realism?*, (Stanford, Calif.: Stanford University Press, 2011).

28. Green, *Japan's Reluctant Realism: Foreign Policy Challenges in an Era of Uncertain Power.*

29. Hughes, "The Democratic Party of Japan's New (But Failing) Grand Security Strategy: From 'Reluctant Realism' to 'Resentful Realism'?," *Journal of Japanese Studies*, vol. 38, no. 1, Winter 2012, pp. 109–140.

PART IV
(RE-)EMERGING WORLD

9 China's National Security Strategy: Waiting at the Crossroads

Kathleen Walsh[1]

Introduction

China's return to regional and likely global power status is well under way. We have a fair, if still incomplete, understanding of China's national interests (or ends), although critical questions remain. At the same time, it is clear that China's military and other elements of national power, be they diplomatic, economic, informational, or other forms of power and influence (i.e., the means by which to secure national interests) are expanding quickly. What we don't have is a clear understanding of the key connective tissue—the grand strategy— that links China's national interests to its growing capabilities and would serve to outline the ways in which China intends to pursue its growing interests with its expanding capabilities.[2] It is not clear, in fact, that China has yet decided on what its national security strategy should be.

China's National Security Interests and Implications for Strategy

Winston Churchill once described Russia as "a riddle wrapped in a mystery, inside an enigma." Speaking in October 1939 and trying to forecast Russian behavior, he added, "but perhaps there is a key. That key is Russian national interest."[3] A similar question is being asked today about China. The fact is that China has yet to make public in a comprehensive official document its national security strategy, which typically begins with a clear outline of a country's core national security interests. In the March/April 2011 volume of *Foreign Affairs*, a leading Chinese academic and policy adviser, Wang Jisi, in fact calls on his government to do just that, noting that "Whether China has any such strategy is open to debate . . . the Chinese government has yet to disclose any document that comprehensively expounds the country's strategic goals and the ways to achieve them."[4] The closest Beijing has come to issuing one publicly is its White Paper on *China's Peaceful Development* issued in September 2011, which outlines China's overarching development strategy, including a broad overview of China's national security outlook, foreign policy, and defense strategy, but offers few specifics.[5] To develop and present a detailed NSS of how China seeks to pursue its interests would be a welcome effort, as continued uncertainty in the strategic domain combined with ever-increasing demonstrations of China's growing powers serves only to expand concerns in the region and beyond of a rising *potential* threat, something Beijing has said it wishes to avoid. In the meantime, we are left to surmise China's actual national security intentions and strategy. As Churchill aptly noted, perhaps the best way to divine what another country's strategy might be is to look to its national interests.

What are China's national interests? Although we do not have the convenience of a formal NSS, there are substantial public statements, a biennially issued *Defense White Paper*, and numerous multiyear, state-issued development strategies and five-year plans that provide rhetorical insights into what China views as its national interests and likely national security strategy. This, and a plethora of academic writings and presentations, gives us material to compare over time and language to parse for strategic nuances but does not provide the clear, unambiguous, and official statement of interests that an NSS serves to outline.

Because national interests are of a vital and enduring nature, they do not tend to vary much over time (as opposed to short-term policy objectives), barring significant changes in a state's internal or external environment. The opposite is also true: significant changes to a state's internal or external environment can have a substantial impact on a state's perceived national interests. China has been undergoing rapid economic and industrial development over the past three decades and vastly expanding its external trade relationships, in terms of both inward and outward investments and exchanges. As such, a change—likely an expansion—in China's national interests can be expected to follow. This, however, makes the lack of a formal NSS from Beijing all the more problematic, raising questions both at home and abroad about what constitutes contemporary Chinese national interests and how far they have expanded along with China's rise. The 2011 *Peaceful Development* White Paper states China's "core interests" as including "state sovereignty, national security, territorial integrity and national reunification, China's political system . . . and overall social stability, and the basic safeguards for ensuring sustainable economic and social development," terminology that is familiar but also subject to interpretation. Moreover, as Wang notes, several factions have emerged in Chinese policy circles as to what strategy would best serve Chinese national interests, each of which takes a very different approach, with starkly distinct implications for China, the region, and the world. In strategic decision-making the ends (interests) inform the ways (strategy) or rationally should, and this factional debate suggests there remains some question about the exact definition of China's national interests. Yet, how China defines its national interests will dictate its NSS and the means employed to pursue these interests.

China's Strategic Assessment and National Security Interests

With its opening to the world and the end of the Cold War, along with decades of fast-paced economic development and military modernization, it is clear that the way in which China views its national interests vis-à-vis the rest of the world today is distinct from how Beijing viewed itself and its role decades ago. While Mao Zedong and Zhou Enlai certainly had global balance of power strategic considerations in mind when aligning in the Cold War first with the Soviet Union then with the United States, China's activities at home and on the global stage were more constrained than conceived in Beijing today. As China's "Open Door" policy began to reap dividends in the 1980s and 1990s, Deng Xiaoping's overarching strategic approach to securing China's national interests remained nonetheless the cautionary "hide and bide" strategy.[6] But spurred on by an increasingly globalized international system that continues to fuel the mainland's economic engine, China's capabilities and reach have grown considerably beyond its own shores, across the region, and into the global arena; so too, it seems, have PRC interests and strategic concerns. In fact, Chinese commentators today at times look to China's dynastic past (i.e., prior to the perceived "century of humiliation" arising from what China views as 19th and 20th century Western imperialism) for strategic insight as well as to the communist PRC's strategic past. But this raises the question of how far, how seriously, with what aim(s), and in what order of priority do current Chinese national interests reach? The core question, to be blunt, is this: What is China willing to sacrifice, fight, and die for and what is it not?

The answer to this most basic of strategic question remains to some extent a mystery. Although China's 2010 *Defense White Paper* (released in spring 2011) provides a summary overview of the PRC's national security concept, questions remain. According to this latest report, China's national security calculations are based on the following assessment of the regional and international environment:

> Looking into the second decade of the 21st century, China will continue to take advantage of this important period of strategic opportunities for national development, apply the Scientific Outlook on Development in depth, persevere on the path of peaceful development, pursue an independent foreign policy of peace and a national defense policy that is defensive in nature, map out both economic development and national defense in a unified manner and, in the process of building a society that is moderately affluent on a general basis, realize the unified goal of building a prosperous country and a strong military . . . In the face of the complex security environment, China will hold high the banner of peace, development and cooperation, adhere to the concepts of overall security, cooperative security and common security, advocate its new security concept based on mutual trust, mutual benefit, equality and cooperation, safeguard political, economic, military, social and information security in an all-round way, and endeavor to foster, together with other countries, an international security environment of peace, stability, equality, mutual trust, cooperation and win-win.[7]

In short, China sees the present day and coming decades as a period of strategic opportunity resulting from expanding global trade and various forms of exchange that are feeding China's economic, scientific, and technological as well as military modernization efforts. China seeks continued global investment and industrial, technological, and scientific cooperation as a means of advancing its own economy and has adopted a foreign policy strategy of promoting the previously noted "win-win" approach and "new security concept" that it hopes will promote continued strategic cooperation as well as appeal to investors and trade partners across the globe.[8] It's unclear, however, how long Beijing perceives this strategic window of opportunity lasting—Chinese leader Jiang Zemin originally envisioned a twenty-year window of opportunity in 2002, but more recent development plans and policy statements suggest it is anticipated to continue through at least 2030 if not 2050. In recent years, however, PRC strategists also perceive rising complexities in the regional and international security environments as well as nontraditional security challenges against which China seeks to hedge.[9] Simultaneously, China is making clear through its public rhetoric and actions that its perceived interests are growing in accordance with its relative gains in global and regional power and influence. None of this is surprising nor necessarily alarming; what is causing concern among many of China's neighbors and beyond, however, is uncertainty over *how far* China's perception of its national interests extends, what distinguishes Chinese security objectives (what China would prefer) from its national interests (what is considered vital), and what the PRC is planning to do (or not do) to secure these interests.[10]

Contributing to such uncertainties are PRC pronouncements, such as in the serial White Papers, that make clear what "peaceful" steps China plans to pursue and what it will oppose others doing. But these documents do not fully outline what interest(s) Beijing considers vital and is willing—or not—to use force (means) to protect nor how much military force it is actually accumulating in support of these interests. Yet Beijing's defense strategy clearly foresees the potential need to use force and seeks to "leap-frog" the modernization of the People's Liberation Army or military in order to be better prepared to do so. China's once attrition-oriented "people's war" strategic construct of a half century ago has evolved over time to the updated, modern-day strategic concepts of "active defense" and "winning local wars under informationalized conditions." The latter imply the possible use of force not entirely nor necessarily defensive (by Western definition) in nature, thus raising questions about China's contemporary national interests and the use of force in support of them, which an NSS would serve to make more clear by outlining and connecting ends, ways, and means.

Changing Notions of Chinese National Security Interests

Despite these uncertainties, Chinese national interests are in some respects straightforward, if only because some interests are universal. Every state since the Westphalian system emerged nearly four centuries ago, for instance, has had as its core national interest the state's continued existence. Therefore one can presume that survival of the sovereign state is a vital Chinese national interest. Yet this normally simple calculation becomes complicated in China's case. First, there is the internal question: what is the nature of the state itself? This is complicated in two respects: institutionally and geographically. Since 1949 when the PRC was established, the institution of state sovereignty and authority has been the Chinese Communist Party (CCP), with the military obliged to serve the Party. The implication, therefore, is that defense of the Party (over and above the state) is a core national interest. Thus, any internal or external threat to the security of the Party could be considered an existential threat and so is likely to elicit the use of force as was arguably the case in 1989 in the Tiananmen Square uprising. Since then, the CCP has sought to adapt to sociopolitical-economic challenges, with Chinese leaders embracing a quasi-capitalist-cum-socialist approach to economic development while at the same time preserving its central authoritative position as the sovereign power in the Chinese system of government. How long this situation can be sustained is anyone's guess.

Geographical or territorial definitions of the Chinese state are also problematic. The PRC has contested areas on its Western frontier, namely Tibet and Xinjiang, which in recent years have witnessed increased internal protests against PRC rule.[11] These tensions are likely to grow as China's economic development shifts farther inland into the central and western provinces, bringing with it Han Chinese traditions, values, policies, and bureaucracy that are often at odds with local customs and ethnic identities. Moreover, while the mainland's formal reintegration of Hong Kong and Macao in 1997 and 1999, respectively, has demonstrated a peaceful path to (from Beijing's perspective) reclaiming sovereign parts of China, Beijing's relationship vis-à-vis Taiwan remains unresolved and uncertain. Beijing's strategy seems clearly aimed at reintegrating Taiwan in a similar fashion to Hong Kong and Macao in terms of building close economic, social, and industrial ties such that political reintegration can follow, but with the looming threat of the use of force in the event of a declaration of independence by Taiwan or an attempt to end the civil war that officially persists despite the long-running *modus vivendi* between the two. While recognizing that Taiwan is historically part of China, the United States and others in the international community would not support reintegration via the use of force.[12] This situation raises two core national interest considerations. First, if Taiwan were attacked (or perceived to be threatened) by a third party, would the mainland see it as a core national interest to come to Taiwan's aid? Would that be welcome and/or viewed as legitimate? Though this scenario seems unlikely in the present-day regional security environment, it is a valid strategic question. The other, more oft-considered concern asks what China is prepared to do in the name of national reunification and territorial integrity that is defined as including Taiwan. In short, having lived so long under the current status quo, in large part due to both militaries' inability to unilaterally alter the situation across the Taiwan Strait (particularly given likely U.S. assistance in some form to Taiwan in the event of a use of force), will the mainland assert this perceived national interest if and when it is militarily able to do so in the name of state sovereignty? It remains to be seen; in the meantime, the two sides continue to become more economically and otherwise integrated. While presently not an immediate security or sovereignty concern, any change in the cross-Strait status quo could potentially provoke wider conflict. China repeatedly has termed the issue of sovereignty over Taiwan as well as Tibet and Xinjiang as "vital" or "core" interests.[13]

Other land boundary issues for China persist as well, such as the still-unresolved dispute with India and Pakistan over areas of Kashmir. Yet since even before the end of the Cold War, China has sought to resolve many of the border disputes on its northern, western, and southwestern borders, reaching agreements with most of its fourteen neighboring states, in part to address China's shifting strategic center of gravity, which due to decades of robust economic growth has moved eastward and seaward. Thus, maintaining China's continental sovereignty and border security remains a strategic national interest; it is also a prerequisite for expanding China's interests into the maritime sphere.

The territorial security question with regard to sovereignty becomes even more complicated when ancient versus modern-day maritime claims, conflicts, and conflicting historical maps make determining what constitutes "China" a geopolitical, strategic diplomatic puzzle. China's nine-dashed or U-shaped line depicting a sovereign claim to the South China Sea according to a circular and map submitted to the UN in May 2009 is reportedly based on historical documentation from the 1940s or, according to some analyses, on 15th to 16th century Ming Dynasty–era maps and is the source of considerable consternation in the region, particularly as Chinese naval and maritime forces have increased their activities—at times in ways perceived as deliberately hostile—in contested maritime areas in the past few years.[14] China's more assertive maritime actions in recent years, in a region where overlapping claims to sovereignty vie for legitimacy and access to large undersea oil and gas deposits are in contention, have raised questions about the PRC's potential use of force (mainly restricted thus far to civil maritime rather than naval forces) in support of its perceived national interests in these resource-rich areas.

Another question arising from China's growing power and possibly expanding interests as well is whether—or the degree to which—China seeks to revise the current regional and/or global order, or whether Beijing will decide to be a more status quo, if rising, power. Recent maritime clashes among regional claimants have led some analysts to suspect the former. Following a rash of regional maritime skirmishes involving Chinese maritime vessels over the past few years, the issue boiled over into dialogue at the 2010 multilateral ARF. U.S. Secretary of State Clinton used the occasion to reinforce the principle of freedom of navigation in the South China Sea and elsewhere as a U.S. national interest in what was widely understood to be a rebuke of China's earlier reported claim to the South China Sea as a "core" national interest (implying possible PRC use of force in protection of this interest and to the exclusion of other states' interests).[15] Since then, the PRC appears to have backed off the "core" interest language at least in formal settings (if it was ever used), adopting subsequently the term "important interest" to characterize China's claims to the South China Sea.[16] But questions remain, and the issue is sure to be revisited in the coming years. Add to this the complication of customary international and maritime law's ongoing transition into contemporary multilateral legal constructs such as the UNCLOS EEZ, and one has a recipe for continued diplomatic discord, at best, over conflicting sovereign claims and interests. But, as a member of UNCLOS, China has sought to reinterpret the definition of the EEZ to extend full sovereign rights (versus exclusive *economic* rights) out to the two hundred nautical mile limit (including in the South China Sea). Whether China's revisionist interpretation will gain substantial international support or not and what exact status China's level of interest is or will be in the South China Sea remains uncertain and will perpetuate strategic concerns in the region despite China's articulated "Good Neighbor" policy toward its Southeast Asian neighbors and beyond.[17]

At the same time, China also has a clear interest in protecting the sea lanes of communication (SLOCs) that constitute strategic channels and chokepoints in terms of China's (and others') maritime access to regional and global trade routes, to the region's resources, and outlets to the Pacific and Indian Oceans. China has made clear that secure access to these SLOCs is a core national interest for both economic and security reasons. But the means by which Beijing intends to secure this interest also remains unclear. Presumably, China's acquisition and deployment of a former Russian aircraft carrier will aid in such an endeavor. But China's strategy for developing and employing such forces remains a topic of much debate both in China and beyond.

As it made clear in the 2011 *Peaceful Development* White Paper, China's continued economic development and modernization clearly remain core national interests for which the PRC requires continued and, ideally, secure access to an ever greater supply of global resources. China's foreign policy approach under Hu Jinto sought explicitly to "safeguard the interests of sovereignty, security, and development." Hu's successor as CCP General Secretary and Central Military Commission Chairman, Xi Jinping, appears thus far to be following a similar approach, holding "firm in safeguarding China's sovereignty, security, and territorial integrity."[18] Scholars inside and outside of China are busy trying to interpret the implications of this new policy formulation. Yet China's economic development and modernization clearly remain a primary focus, as it has been for more than thirty years, with Xi renewing the country's adherence to developing "Socialism with Chinese characteristics" and calling for a "great rejuvenation of the Chinese nation."[19] China today is also in direct, global competition for strategic resources with Western, neighboring, and other developing powers. As such, China's development interests exceed its national boundaries and are increasingly global in scale and so would certainly comprise part of any Chinese NSS.

China's decades-long and rapid-paced development strategy has been encumbered, however, by internal and external challenges, particularly in recent years. Internally, China faces challenges to "social harmony" due to imbalanced growth rates between rich and poor and between coastal and inland areas, from rampant corruption, environmental degradation, and calls for greater political rights, all of which Chinese leaders seek to address in current national development plans and strategies. But once again, to the extent that any of this disharmony is perceived to be an existential threat to the CCP, it could lead to a resort to force, as witnessed in 1989. Accordingly, the years since Tiananmen Square have seen the development of more paramilitary forces designed to deal more effectively with internal security concerns while freeing the PLA to focus more—though not exclusively—on external security matters.

In the meantime, China has extended its strategic and defensive perimeter eastward toward the sea as its industrial center of gravity has shifted from the inland to coastal areas. Although many of the early PRC "Third Front" defense industries remain located in the central and western or remote northeastern provinces, much of the high-tech, "informatized" sector of China's economy resides in the large coastal cities of Beijing, Tianjin, Shanghai, Guangzhou, Shenzhen, and areas in between. Thus, rather than draw any potential aggressors into China's interior to be defeated via attrition as in the days of Mao, Beijing's strategic concept today seeks to expand China's defense perimeter eastward and seaward so as to provide a maritime buffer zone to protect the increasingly valuable yet also vulnerable coastal areas. This buffer zone objective contributes in part to China's evolving strategic concepts of expanding its forces and capabilities into what it terms its "near seas" and possibly "far seas" (*jin hai* and *yuan hai*, respectively) and for which China is expanding its civil maritime patrol forces as well as naval capabilities.[20] To the east, south, and seaward, however, is an already crowded zone of state, allied, and international interests of a territorial, maritime, and strategic nature. As noted, expanding Chinese state interests—and projection of forces—into such an environment, even if understandable in terms of a rising power's expanding interests, has and is likely to continue to spark questions and concerns among other parties with interests in the region. Following a rash of regional maritime skirmishes and perhaps aimed at addressing these complexities, Beijing is reportedly in the midst of developing a national maritime strategy.[21]

As China seeks to leverage the strategic window of opportunity it perceives and to reap development opportunities made possible through globalization, Beijing is expanding its overseas investments and sending personnel abroad in search of more resources, technological, and innovative knowledge in order to achieve "indigenous innovation" goals, and pursuing so-called win-win agreements with foreign trade partners, in an effort to expand what Beijing calls Comprehensive National Power. China's reach and interests have expanded accordingly. Chinese students populate more overseas universities than ever before; Chinese enterprises and investors are expanding overseas in both developed and developing countries; and PRC diplomats are engaged worldwide in enhancing China's image, political influence, and soft power abroad by promoting Chinese language resources, Confucian centers, student and faculty exchange programs, infrastructure and resource projects, arms trade, humanitarian relief aid, deployment of peacekeeping forces, and more. Today China is engaged around the globe, politically, economically, and increasingly militarily. The latter is primarily in noncombat roles such as naval port visits, counter-piracy operations in the Gulf of Aden, UN peacekeeping operations (e.g., Haiti, Lebanon, Timor Leste), as well as noncombatant evacuation operations in Libya. These more far-flung, non-traditional "new historic missions" and military operations have been in service of protecting China's national and increasingly global economic interests yet dually provide opportunities to demonstrate China's newly developed maritime, air, logistics, and military personnel capabilities.[22] This expanding global presence raises the question, once again, of what exactly constitutes China's national security interests today, and whether these will keep expanding in terms of both capabilities and areas of interest as China's power and reach grows or whether there is some basis for strategic restraint. Prior policy restraints, such as the principle of noninterference in others' internal affairs, have come into question or, at minimum, have been stretched in recent years as China has participated in more nontraditional overseas activities, such as the aforementioned counter-piracy and peacekeeping efforts, that rely on international authorities at times superseding traditional sovereign claims to noninference by outside actors.[23] Paradoxically, as China accedes to contemporary international norms, it puts in question some of what has traditionally guided China's strategic decision-making and raises questions about what limits, if any, bound China's interests and strategy today.

Another traditional national interest for any state is an interest in protecting one's allies and strategic partners in the region. China's strategic partners in the region arguably include North Korea, Russia (although this continues to be an on-and-off again relationship), and Pakistan. Until recently, Burma could also be included in this list, given still-close military and economic ties. But the Burmese government's recent political reforms and outreach to the United States puts the strength of future Sino-Burmese security relations in question.

China perceives a looming threat from the expanded U.S. presence in South Asia and in the Asia-Pacific theater, especially since the 9/11 attacks, which led to a more than decade-long U.S. and NATO presence in Afghanistan as well as expanded and ongoing counterterrorism and counter-proliferation activities in the region. U.S. foreign and security policies in the past several years also have emphasized a return to a more strategic focus on the Asia-Pacific region, which was formalized in a new 2012 "pivot" or "rebalancing" strategy announced by the White House and Pentagon and aimed at enhancing the network of U.S. alliances and partnerships in the region.[24] China's response has been to expand its own bilateral and multilateral relations with states in the region, as evidenced in the more expansive activities undertaken by the SCO, regular involvement in the East Asia Summit, and other regional groupings, as well as through growing cooperative programs or investments in allied and neighboring economies and infrastructures, particularly in rail, pipeline, and port facilities. China also has expanded its military partnerships, arms sales, and exercises.[25] At the same time, China has been reluctant to join Western countries and regional powers in consistently condemning continued North Korean provocations, although North Korea's antics often run counter to China's own presumed interests in preserving a reasonably stable regional environment conducive to continued economic growth and foreign investment. For the PRC, North Korea's continued existence, though problematic at the very least, provides a buffer zone against a U.S. allied presence directly on China's border. This is preferable to a reunified Korean peninsula more closely allied with the United States than the PRC. China's relations with Burma, Russia, and Pakistan remain complicated as well, given each country's problematic relations with the West. The United States' courting of India in recent years has also added complexity to China's strategic calculus.

Two other domains that China perceives as sovereign have emerged in recent years: space and cyberspace. China's scientific and technological advances in its lunar, satellite, and other space programs have brought the PRC into the elite club of space-faring nations. With this ability comes concern over space assets and the potential for threats to or attacks on these national resources. Although China has been a long-time proponent of the non-weaponization of space at the UNCD, this position historically has posed minimal political risk given the unlikely shift in position by the United States and other parties dependent on space assets for military capabilities. This position was further undermined in 2007 when China tested its own ASAT weapon. As China's space and satellite capabilities and inventories grow in expectation of "local wars under informationalized conditions," it is likely, though not certain, that Beijing's national interests in securing these assets on sovereign grounds will grow as well.[26]

Cyberspace, too, poses paradoxical security and sovereign concerns for China. The PRC disputes the normative U.S. view on cyberspace—that the Internet should remain a transnational rather than a national asset for the sake of freedom of speech and information, as outlined most forcefully in January 2010 by U.S. Secretary of State Hillary Clinton—and seeks to maintain its tight, national network security protocols in order to prevent what it views as external interference in its domestic affairs.[27] At the same time, the repeated and growing incidents of international computer hacking that are traced back to Chinese sources puts its position on cyber sovereignty in question. Most importantly, it is unclear in either domain whether or under what circumstances perceived threats to China's space or cyberspace capabilities would actually elicit a Chinese use of force or not, an issue the United States and others are also struggling to define.

Finally, having early on determined a national interest in developing a nuclear capability to guard against external, existential nuclear threats, the PRC attained nuclear weapon status in 1964 and is believed to have maintained a minimal nuclear deterrent posture since then. But with the advent in the West of precision guided munitions and conventional weapons of a similar destructive magnitude, as well as a U.S. interest in developing and deploying ballistic missile defenses, China's strategic calculus of its national interests appears to have shifted in tandem. Public sources relate that China is undergoing a substantial nuclear upgrade, likely by employing more advanced, mobile missiles with modernized, multiple warheads, expanding its nuclear submarine fleet, overall nuclear weapons stockpiles, and possibly even a BMD system or related technologies of its own. But it is unclear exactly how this modernization is occurring, and at a time when U.S.–Russian arms control measures have cut deeper into strategic inventories, thus further raising strategic uncertainty.

China's National Interests: Expanding but How Far?

How Beijing pursues these objectives and which will be considered (or not) to be *vital national interests* is still unclear. The approach outlined in the *Defense White Papers* calls for peaceful, mutually beneficial resolutions to such conflicts, which indeed has been an approach adopted in several instances over the past decade (e.g., resolution of land-based border disputes with most neighboring states and the signing of a 2002 Code of Conduct agreement with ASEAN partners to shelve maritime disputes).[28] Yet there have also been growing instances over the same period of time of provocative Chinese acts and statements advocating broad-reaching PRC interests, making for greater uncertainty as to ultimate Chinese aims, strategic intent, and means employed, particularly in terms of the potential use of force in the future. As with the questions over Taiwan, does a more accommodating approach reflect China's enduring national interests or is it serving China's objective *at this time* and subject to change as the Mainland develops a greater capability to defend its expanded national interests? It can be argued both ways, but only Beijing can answer authoritatively.

Reflections: China's National Interests

Based on the foregoing discussion, it is possible to outline a reasonable estimation of China's perceived national interests and from this forecast what China's subsequent national security strategy and potential force planning might emphasize.[29]

- Maintain CCP rule and political stability: Defend state sovereignty against enemies foreign and domestic; and maintain stability through "social harmony."
- Secure territorial sovereignty and defense of China: Ensure a deterrent against assured destruction; defend territorial claims; extend the defense perimeter from western frontier regions to coastal areas and maritime claims; develop a maritime buffer zone; and protect space and cyberspace domains.
- Sustain the engine of growth with economic and science and technology development as the main focus: Attract regional and global investment and expand economic interdependence in an effort to integrate territories and attain an indigenous innovation capability; leverage globalization and the strategic window of opportunity to serve continued modernization efforts in economic, scientific, and technological as well as military spheres; pursue peaceful development, cooperation and win-win solutions in economic and strategic relations; secure access to and supply of resources necessary for China's continued modernization (i.e., SLOCs); and build "comprehensive national power" (i.e., expand hard and soft powers).
- Maintain against external threats the sovereignty and security of China's allies and partners in the region.

This list is reflective of recent statements emanating from Chinese media (some officially sanctioned, some not) as well as of PRC activities in the past few years.[30] Presuming this is more or less an accurate depiction of China's actual core national interests, what sort of national security strategy and force planning would such a list logically imply?

Taking this somewhat narrow list of national interests—and leaving aside aforementioned objectives that, though important, fall short of what are likely considered vital interests—a Chinese national security strategy and the type of forces necessary to secure these interests would clearly dictate a modern and robust military and nuclear capability intended to secure China's Party leadership as well as China's territorial, space, and cyberspace domains. It would include weapons platforms and capabilities designed to expand China's maritime buffer zone so as to offer greater strategic depth and protect the rich and vital coastal economic zones and sovereign territories. It would add enhanced force projection and logistics capabilities (e.g., ships, aircraft, and logistics) to protect vital SLOCs, and it would include numerous diplomatic and soft power capabilities used to promote and secure China's interests overseas. In fact China has pursued all of these capabilities and more. The "more," however, is where China's lack of transparency on military and strategic affairs and perceived overreach in articulating strategic interests or objectives is causing concerns across the region. Beijing states that it is pursuing a "peaceful development" strategy during a strategic window of opportunity, but what does this mean if or when China's interests clash with others', when two or more of its own core interests are at odds, or if Beijing's perception of the strategic window of opportunity changes? A publicly articulated national security strategy could serve to answer these and related questions.

China's claim to nearly all of the South China Sea, for instance, and recent skirmishes there and elsewhere in the region over sovereign territorial and resource claims are based on contested historic and legal bases, heightening security concerns among China's neighbors and conflicting with China's stated interest in continued regional and global stability for the sake of economic, technological, and military modernization. But are such territorial claims to sovereignty fundamentally essential to securing China's vital interests in SLOC security or resource supply in this area? This is debatable, as other potential means of securing these interests exist, such as a regional code of conduct. Arguments positing that China's and other claimants' interests in this region are threatened by what others are doing and what resources they are extracting in the region also suggest that time and opportunity to exploit the region's natural gas and oil resources are the key variables of national interest, more so perhaps than promoting principled stakes in territorial sovereignty.[31] While more cooperative ways and means might not be Beijing's preferred approach, they are among the conceivable ways in which China could meet its national interests in securing access to regional SLOCs and resources. If China decides on a more assertive strategy, this will raise questions about the extent of China's national interests and the forces it plans to employ in order to secure them.

Likewise, China's newly commissioned aircraft carrier (and possible sizeable fleet to follow) begs the question of what China aims to use it for, since carriers can project force over long distances, suggesting China is pursuing greater regional and global interests than Beijing's estimation of its current security situation would necessarily require. The issue of Taiwan and the potential for a U.S.–China clash is the most obvious carrier rationale, sparking renewed concerns over China's potential use of force (which Beijing has long declined to renounce in this instance), and yet China's carrier ambitions appear to be more wide ranging. If so, this again poses uncertainty over China's own perception of its vital or core interests as distinguished from its long-term and important objectives or aspirations but not-quite-vital national and international security interests. If China's carrier is intended to counter the United States' presence in the region, as its anti-access-oriented defense strategy appears designed to do, then does this suggest a fundamentally revisionist strategic outlook on the part of Beijing? Or might China's pursuit of a carrier force serve to secure other national interests while allowing a rising state to pursue a more balanced distribution of power in the region, but in cooperation with a still-dominant or perhaps peer yet close strategic partner? Having a sense of China's national interests and its increasing force capabilities without a clear outline of the ways in which it seeks to secure its interests and employ its forces (i.e., as outlined in an NSS) is likely to promote worse-case scenario planning and heightens the potential for conflict.

Thus, it appears that in terms of determining an NSS, China is at a crossroads. The further the PRC pursues its aspirational strategic and military objectives, the more it risks undermining its vital national interests in preserving the peaceful window of opportunity for economic, science and technology, and military modernization needed to secure these interests. As a rising power, China is expected to embrace a greater set of obligations and interests, but not at the expense of other parties' sovereign and security interests or through the use of force; such is the basic premise underlying the notion of welcoming China's rise as a responsible stakeholder in the international system. The ambiguity surrounding China's expanding interests, however, will continue to feed insecurity and uncertainty, as would a formal declaration by Beijing of a set of national interests and/or national security strategy that conflicts with existing norms (as indicated in the sharp and rapid response to reports of China's supposed claim in 2010 to the South China Sea as a "core" interest).[32] So what is a rising and strategically aspiring country to do?

In advocating for a comprehensive Chinese national security strategy, Wang Jisi outlines three contemporary and distinct schools of thought in China on this question: 1. treat the United States as an enemy; 2. enhance China's power and influence via strategic alliances and expanded participation in nontraditional security matters (e.g., counterterrorism activities, antipiracy operations, UN peacekeeping, and so on); and 3. continuation of Deng Xiaoping's admonition of "biding one's time by keeping a low profile in international affairs."[33] The difficulty for foreign analysts is that aspects of each of these approaches is evident in Chinese foreign policy pronouncements as well as diplomatic and military activities, making an assessment of which one of these—or mix of these approaches—is guiding, will guide, or will prevail as an official Chinese national security strategy very difficult. Moreover, because each has distinct strategic implications for China's neighbors, partners, and potential rivals, this has led some—particularly the United States and its allies in the Asia-Pacific—to increasingly hedge against a possible worst-case scenario of China as revisionist power or aggressor while still seeking cooperative, if competitive, opportunities to engage.

Where a good number of China's sovereign claims are in question or in dispute, it raises the specter of strategic miscalculation and possible escalation of small-scale skirmishes into something more strategically significant. Wang cites such dangers as a reason for Beijing to publish a formal and comprehensive grand strategy, advocating one that avoids the risks of the first two schools of thought but provides a more balanced approach to enhancing China's hard and soft powers as a means of securing the PRC's more well-defined national core interests. In order to counter the concerns the latter approach raises, Wang advises, among other efforts, greater transparency by China in the military sphere and enhanced PRC cooperation in regional and international regimes and security dialogues. In fact, China and the United States engage in more than sixty formal dialogues each year, yet strategic questions persist.[34] More dialogue, therefore, would be welcome by the international community and could help inform, if not entirely alleviate, such concerns. In the meantime, neighboring states and regional powers will continue to hedge and be wary of aspects of Chinese military modernization that appear to exceed China's apparent core national security interests.

China's national security strategy remains a work in progress as Beijing seeks to expand its reach, protect its expanding interests, manage its continuing rapid-paced rise and development, and determine what interests and objectives it will and will not, respectively, use military force or other means to secure. China's newly designated leader, Xi Jinping, has declared that China stands "firm in safeguarding China's sovereignty, security, and territorial integrity," but how this interest will be interpreted into grand strategy and force planning is as yet unclear. Given the growing contradictions between China's apparent strategic ends, ways, and means, the lack of a formal, declared national security strategy will continue to be a key source of strategic uncertainty.

Notes

1. The views presented herein are those of the author alone and in no way represent the views of the U.S. Government, U.S. Navy or Naval War College.

2. This chapter takes a rational actor-style approach to analyzing China's NSS given that 1. China has not published an NSS as yet, requiring analysts as a first step to determine what the NSS might look like based on an assessment of China's national interests; 2. the purpose of an NSS, ideally, is to outline in logical fashion the *ways* in which a country's interests (*ends*) are linked to its *means*, a logic that the RA perspective can help illuminate; and 3. there is not space available here to delve into the kind of highly detailed analysis that is required to conduct domestic levels of analysis such as governmental politics among China's elite nor of organizational behavior among China's foreign policy bureaucracy, although these analytical approaches can yield useful policy decision-making insights as well.

3. The full quote, according to Bartlett's Familiar Quotations, is "I cannot forecast to you the action of Russia. It is a riddle, wrapped in a mystery, inside an enigma; but perhaps there is a key. That key is Russian national interest."

4. Wang is a well-known, respected scholar of U.S.–China relations and Dean of the School of International Studies at Peking University in Beijing. See Wang Jisi, "China's Search for a Grand Strategy: A Rising Great Power Finds Its Way," *Foreign Affairs*, vol. 90, no. 2 (March / April 2011), pp. 68–79; p. 1.

5. Information Office of the State Council, China's Peaceful Development (September 6, 2011).

6. Deng's famously advised: "Coolly observe, calmly deal with things, hold your position, hide your capabilities, bide your time, accomplish things where possible."

7. China's 2008 Defense White Paper (DWP) promoted similar themes. See Information Office of the Chinese State Council, *China's National Defense in 2008* (Beijing, January 2009).

8. The 2010 DWP explains that "By connecting the fundamental interests of the Chinese people with the common interests of other peoples around the globe, connecting China's development with that of the world, and connecting China's security with world peace, China strives to build, through its peaceful development, a harmonious world of lasting peace and common prosperity." See Information Office of the Chinese State Council, *China's National Defense in 2010* (Beijing, March 2011). This sentiment aligns with China's appeal in recent years to a "new security concept" for the post–Cold War world of "mutual trust, mutual benefit, equality and coordination." See Xia Liping, "The New Security Concept in China's New Thinking of International Security," *International Review*, vol. 34 (Spring 2004), pp. 29–42.

9. Notably, this is a subtle change from the 1990s, when Chinese strategic rhetoric described the external environment as being overwhelmingly positive. The 2010 DWP, in contrast, notes that "Asia-Pacific security is becoming more intricate and volatile. Regional pressure points drag on and without solution in sight. There is intermittent tension on the Korean Peninsula. The security situation in Afghanistan remains serious. Political turbulence persists in some countries. Ethnic and religious discords are evident. Disputes over territorial and maritime rights and interests flare up occasionally. And terrorist, separatist and extremist activities run amok. Profound changes are taking shape in the Asia-Pacific strategic landscape. Relevant major powers are increasing their strategic investment. The United States is reinforcing its regional military alliances, and increasing its involvement in regional security affairs," adding that "China is still in the period of important strategic opportunities for its development, and the overall security environment for it remains favorable."

10. More expansive discussions on the question of China's security strategy and contemporary national interests can be found in, for instance, Avery Goldstein, *Rising to the Challenge: China's Grand Strategy and International Security* (Stanford, Calif.: Stanford University Press, 2005); *China's Ascent: Power, Security, and the Future of International Politics*, Robert S. Ross and Zhu Feng, eds. (Ithaca, N.Y.: Cornell University Press, 2008); and Hugh White, *The China Choice: Why America Should Share Power* (Australia: Black Inc., 2012).

11. Xinjiang came under PRC control in 1949; Tibet was annexed in 1950. Both are classified as administrative autonomous regions.

12. As outlined by the U.S. State Department, U.S. policy on Taiwan is clear on this point: "On January 1, 1979, the United States changed its diplomatic recognition from Taipei to Beijing. In the U.S.-P.R.C. Joint Communiqué that announced the change, the United States recognized the Government of the People's Republic of China as the

sole legal government of China and acknowledged the Chinese position that there is but one China and Taiwan is part of China . . . The United States position on Taiwan is reflected in the Three Communiqués and the Taiwan Relations Act (TRA). The U.S. insists on the peaceful resolution of cross-Strait differences and encourages dialogue to help advance such an outcome. The U.S. does not support Taiwan independence."

13. Da Wei, "A Clear Signal of 'Core Interests' to the World," *China Daily* (August 2, 2010); Edward Wong, "China Hedges Over Whether South China Sea Is a 'Core Interest' Worth War," *The New York Times* (March 31, 2011), p. A12; and Michael D. Swaine, "China's Assertive Behavior Part One: On 'Core Interests,'" *China Leadership Monitor*, no. 34 (Palo Alto, Calif.: Hoover Institution, February 22, 2011).

14. Li Jinming and Li Dexia, "The Dotted Line on the Chinese Map of the South China Sea," *Ocean Development & International Law*, vol. 34 (London: Taylor & Francis Inc., 2003), pp. 287–295.

15. China reportedly claimed the South China Sea as a "core" national interest in spring 2010. See Mark Landler, "Offering to Aid Talks, U.S. Challenges China on Disputed Islands," *New York Times* (July 23, 2010), p. A4. For a more detailed and contrasting view, see Swaine (2011).

16. Professor Su Hao (China Foreign Affairs University in Beijing) commenting on "Interests and Positions of Parties in the South China Sea," at the conference on "Maritime Security in the South China Sea," hosted by the Center for Strategic and International Studies (CSIS) in Washington, D.C. on June 20, 2011.

17. Xia Liping, "The New Security Concept in China's New Thinking of International Security," *International Review*, vol. 34 (Spring 2004), pp. 29–42.

18. Xi Jinping was expected to formally succeed Hu Jintao as Chinese president at the March 2013 Party Congress.

19. Statement by Xi Jinping at a CCP Central Committee Workshop for new National People's Congress Members. See "Xi Stresses Socialism with Chinese Characteristics," China.org.cn (January 6, 2013), accessed online at http://www.china.org.cn/china/2013-01/06/content_27596486.htm.

20. On China's expanding civil maritime security forces see "China to Strengthen Maritime Forces Amid Disputes," *People's Daily Online* (June 17, 2011), and on China's expanding naval strategy, see Nan Li, "China's Naval Modernization: Causes for Storm Warning?," presentation before the 2010 Pacific Symposium, hosted by the Institute for National Strategic Studies (INSS) of the National Defense University (Washington, D.C.: INSS/NDU, June 16, 2010).

21. China apparently has a maritime strategy document, according to more than one Chinese interlocutor. Yet, in line with other contemporary observations, a recent commentary notes that "The military elites in China are calling for a more extensive national maritime strategy, which is under review." See Yang Fang, "China's New Marine Interests: Implications for Southeast Asia," *RSIS Commentaries*, no. 9/7/2011 (July 4, 2011).

22. China's "new historic missions" for the PLA were codified in 2007 as an amendment to China's Constitution. They are intended to align party and military objectives, namely to "provide an important guarantee of strength for the party to

consolidate its ruling position; provide a strong security guarantee for safeguarding the period of strategic opportunity for national development; provide a powerful strategic support for safeguarding national interests; [and] play an important role in safeguarding world peace and promoting common development." See Office of the Secretary of Defense, *Military and Security Developments Involving the People's Republic of China 2010* (Washington, D.C.: U.S. Department of Defense, 2010), pp. 18–19.

23. It should be noted that in such instances where Chinese personnel have been involved that Beijing has insisted on international legal authorities and has sought to gain consent from the government(s) in question prior to PRC personnel participation.

24. U.S. Department of Defense, "Sustaining U.S. Global Leadership: Priorities for 21st Century Defense" (January 2012).

25. In recent years, China has conducted numerous joint exercises in the region and elsewhere with, for instance, Russia, Pakistan, Gabon, Singapore, Mongolia, and fellow Shanghai Cooperation Organization members. Some analysts suggest China might also be seeking to expand its strategic presence and influence in the region. Media reports suggest establishment of a joint China-Burma naval unit, while rumors persist of a possible Chinese interest in establishing a naval base at the Gwadar port in Pakistan and/or elsewhere in the region. Thus far, the latter "string of pearls" theory advocating a PRC interest in establishing regional military bases has yet to be corroborated. See "China to Set Up Naval Unit in Myanmar," *New Delhi Political and Defence Weekly* (January 11–17, 2011); and Urmila Venugopalan, "Pakistan's Black Pearl: The Hype about a Chinese-built Port on the Arabian Sea Says More About Islamabad's Desperation than It Does about Beijing's Imperial Ambitions," *Foreign Policy* (June 3, 2011).

26. On China's space and weaponization strategy prior to the 2007 ASAT test, see Zhang Hui, "Space Weaponization and Space Security: A Chinese Perspective," *China Security*, issue 2 (2006).

27. Secretary of State Hillary Rodham Clinton, "Remarks on Internet Freedom," presented at the Newseum, Washington, D.C. (January 21, 2010), available online at http://www.state.gov/secretary/rm/2010/01/135519.htm.

28. Association of Southeast Asian Nations (ASEAN), "Declaration on the Conduct of Parties in the South China Sea," available online at http://www.aseansec.org/13163.htm.

29. The issue of what constitutes China's "core interests" has become quite contentious in the past few years and became particularly inflamed when influential foreign policy advisor State Councilor Dai Bingguo *reportedly* proclaimed China's "core interests" at the 2010 U.S.–China Strategic and Economic Dialogue to include the South China Sea. A year earlier, however, Dai's definition of core interests included simply the need to "uphold our basic systems, our national security; and secondly, the sovereignty and territorial integrity; and thirdly, economic and social sustained development." According to Wang Jisi, Dai Bingguo further defined core interests in an article last December as follows: "first, China's political stability, namely, the stability of the CCP leadership and of the socialist system; second, sovereign security, territorial integrity, and national unification; and third, China's sustainable economic and social development."

Similarly, a 2010 study cites China's fundamental national security interests as "... first, to defend national sovereignty and territorial integrity, to create a conducive and peaceful development of the reform of the external environment; Second, as a practical task to achieve reunification of the motherland; third is to maintain security and stability of civil society; four is to protect the country in its rightful place on the international stage, as opposed to hegemony, to contribute to safeguarding world peace." On the latter, see Guo Ji, Zhang Guangzhong, and Zhao Pan, "China Military Strategies in the New Military Transformation Times," *Value Engineering* (2010), pp. 135–136; see also Wang (2011); and U.S. Department of State, "Closing Remarks for U.S.–China Strategic and Economic Dialogue," SED2009, transcript (July 28, 2009).

30. Wang cites a more conservative list of "core interests." He states, "Apart from the issue of Taiwan, which Beijing considers to be an integral part of China's territory, the Chinese government has never officially identified any single foreign policy issue as one of the country's core interests." He notes that "some Chinese commentators reportedly referred to the South China Sea and North Korea as such, but these reckless statements, made with no official authorization, created a great deal of confusion. In fact, for the central government, sovereignty, security, and development all continue to be China's main goals. As long as no grave danger—for example, Taiwan's formal secession—threatens the CCP leadership or China's unity, Beijing will remain preoccupied with the country's economic and social development, including in its foreign policy." Wang (2011).

31. See, for instance, Li Mingjiang, "Reconciliing Assertiveness and Cooperation? China's Changing Approach to the South China Sea Dispute," *Security Challenges*, vol. 6, no. 2 (Winter 2010), pp. 49–68. Li argues that "As is true for other parties, energy resources in the South China Sea are perhaps the most important attraction for China. Different from other claimant states, perhaps, is the strategic importance of oil and gas in the area: Chinese analysts view the natural resources in the South China Sea as a critical requirement for the future of China's national economy... In light of this, it is no surprise that China has been quite upset by the exploitation of energy resources in the South China Sea by other claimant countries" (p. 51).

32. According to research by Michael Swaine, and despite media reports to the contrary, China apparently did not—at least not in a formal or public manner—claim the South China Sea as a "core" interest during the 2010 S&ED dialogue. Nonetheless, the United States responded as if it had (with many U.S. officials believing so), leading to much discussion, debate, and consternation in the region. Swaine (2011).

33. Other scholars identify a wider range of cross-cutting schools of thought. See, for instance, David Shambaugh, "Coping with a Conflicted China," *The Washington Quarterly*, vol. 34, no. 1 (Winter 2011), pp. 7–27, which posits a spectrum of views on China's role in international affairs ranging from nativist to globalist but finding a preponderance of realist views.

34. Kenneth Lieberthal and Wang Jisi, *Addressing U.S.–China Strategic Distrust*, John L. Thornton China Center Monograph Series No. 4 (Washington, D.C.: The Brookings Institution, March 2012), p. 1.

10 India: Security Policy in a Strategic Void

Harsh V. Pant

By all accounts, India is on a self-sustaining trajectory of economic growth. This may seem hard to believe, given that its economy was in such dire straits in 1991 that the government was forced to pawn 67 tons of gold to the Bank of England and the Union Bank of Switzerland in order to shore up its dwindling foreign exchange reserves of a measly $2 billion.[1] The distance that the country's economy has traveled since then could be gauged by the fact that in November 2009, with the nation's foreign exchange reserves standing at US$285 billion, India decided to buy 200 tons of gold from the IMF.[2]

The financial crisis of 2008–2009 did not impact India as severely as it did some of the countries in the West. In that period, the country experienced slower exports and lost some liquidity, but overall its economy has weathered the storm and remains highly resilient. Like other emerging markets, India appears to be recovering far faster than developed nations. The country's recovery from the global financial crisis has also been broad-based, with its three main sectors—industry, agriculture, and services—all performing well. As a result, the economy is growing and levels of savings, investment, and foreign exchange reserves are all high.

Though there has been a slowdown in the rate of economic growth primarily because of domestic political turmoil, there are few who question the ability of the Indian economy to continue growing in the near future. As regards the stimulus behind this growth pattern, the World Bank argues that as a net importer of goods India has sustained growth by utilizing domestic demand. In contrast, export-based economies, such as China's, are far more dependent on global consumer demand, so the bank suggested that India's growth rate might even outpace that of China.[3] This would make India's economy the fastest growing large economy in the world. In addition, its position might also enable it to benefit from increased capital investment and clout as a result of the slowing of growth in the East Asian region.

If the global balance of power is indeed shifting from the Atlantic to the Pacific, then the rise of India, along with China, is clearly the indisputable reality that few can dare to dismiss any longer. As a consequence, India is now being invited to the G8 summits, is being called upon to shoulder global responsibilities from the challenges of nuclear proliferation to the instability in the Persian Gulf, and is increasingly being viewed as much more than a mere South Asian power. After decades of marginalization due to the vagaries of the Cold War, its own obsolescent model of economic management, and the seemingly never-ending tensions with Pakistan, India is finally coming into its own with a self-confidence that comes with growing capabilities. Its global and regional ambitions are rising, and it is showing aggressiveness in its foreign policy that had not been its forte before.[4] There is clearly an appreciation in the Indian policy-making circles of India's rising capabilities. It is reflected in a gradual expansion of Indian foreign policy activity in recent years, in India's attempt to reshape its defense forces, and in India's desire to seek greater global influence.

This chapter examines the factors that are shaping the national security strategy of India as it takes center stage in global politics after decades of being on its periphery. It argues that although there is an emerging consensus among Indian policy-makers and the larger strategic community that the old foreign policy and national security framework, perhaps adequate for the times when it was developed, is no longer capable of meeting the challenges of the times, there is little agreement on a strategic framework around which India should structure its external relations in the present global context.

Structural Change and Its Aftermath

The transformation in Indian foreign and security policies since the end of the Cold War has been the result of a number of factors. All through the Cold War years, India saw itself as the leader of the Third World even as the Third

World group of states existed more in myth than in reality. While the idea of a nonaligned foreign policy may have been devised to prevent Indian foreign policy from becoming hostage to the Cold War rivalry between the United States and the former Soviet Union, in practical terms it led to a certain ideological rigidity that prevented India from protecting and enhancing its vital interests in an anarchical international environment. As the Cold War drew to a close, India was forced to reorient its economic and foreign policies to the changing global realities and in less than two decades seems on the cusp of achieving the status of a "great power."[5]

While some were proclaiming the end of history with the fall of the Berlin Wall, in many ways it was the beginning of history for Indian foreign policy, freed as it was from the structural constraints of a bipolar world order. It lost its political, diplomatic, and military ally with the demise of the Soviet Union, and its economy was on the threshold of bankruptcy. There was domestic political uncertainty, with weak governments unable to last for a full five-year term as a plethora of internal security challenges became more prominent. The ignominy of having to physically lift bullion to obtain credit pushed India against the ropes, and the national psyche was at its most vulnerable. It was against this background that the minority government of the late P. V. Narasimha Rao, which came to office in 1991, had to formulate its economic and foreign policy to preserve Indian interests in a radically new global environment. And slowly but surely, the process continues to unfold as India has tried to redefine its place in the international system in consonance with its existing and potential power capabilities. Michael Mandelbaum has argued:

> [s]imilar security policies recur throughout history and across the international system in states that, whatever their differences, occupy similar positions in the international system . . . The security policies of very strong states are different from those of very weak ones, and both differ from those of states that are neither very strong nor very weak."[6]

Structural constraints, in other words, force states toward a particular set of foreign policies in line with their relative position in the international system. And as that position undergoes a change, so will the foreign policy of that state change. As a nation's weight in the global balance of power rises, it becomes imperative to pay greater attention to the systemic constraints. Rising states, suggests Christopher Layne, have "choices about whether to become great powers. However, a state's freedom to choose whether to become a great power is in reality tightly constrained by structural factors. Eligible states that fail to attain great power status are predictably punished. If policy-makers of eligible states are socialized to the international system's constraints, they understand that attaining great power status is a pre-requisite if their states are be secure and autonomous."[7] While ignoring the structural imperatives carried little cost when India was on the periphery of global politics, it has been suggested that this can have grave consequences now when Indian capabilities have risen to a point where India seems poised to play a significant role in global politics.[8]

Nonalignment and Beyond

India's first Prime Minister, Jawaharlal Nehru, was certain, well before two blocs appeared on the global scene in 1947, that "the other countries could not tolerate that the rich prize of India should fall again to another power."[9] This meant that India needed to be friends with all the major powers after 1947, that is, with both blocs. At the same time, Nehru also argued that India would have to "plough a lonely furrow" at a time of the Cold War between the two superpowers, thereby defending nonalignment as "the only honourable and right position" for India.[10] As a result, India's policy of nonalignment emerged, which reasoned that Indian interests would best be served by retaining autonomy from the two sparring blocs, a strategy that ostensibly gave India a stature and an influence well beyond its economic and military weight. But there were inherent problems with this worldview, as the quest for regional ascendancy was coupled with the quest for great power status. India's major power status was taken for granted in the conduct of the nation's foreign policy without any concern about the regional or global balance of power. India was a third-world country with a first world attitude that did keep a sense of independence and resistance to imperialism alive, even as it did not allow India to leverage the power of small nations to further its own goals.[11]

Some have defined this proclivity as a "mini-state syndrome," those states that lack the material capabilities to make a difference to the outcomes at the international level and often denounce the concept of power in foreign policy-making.[12] India had long been such a state, viewing itself as an object of the foreign policies of powerful nations. Consequently, the Indian political and strategic elites developed a suspicion of power politics, with the word "power" itself acquiring a pejorative connotation regarding foreign policy. The relationship between power and foreign policy was never fully understood, leading to a progressive loss in India's ability to wield power effectively in the international realm. It has been suggested that today when India wants to shape the international system, it is more important than ever that its foreign policy is "anchored on a planned augmentation of the power of the nation as a whole."[13] Even the pious declarations of world peace, disarmament, and global development that India propounds on the world stage will be taken seriously only if they come from a state that the international community perceives as having the will and the ability to convert its rhetoric into reality.

The biggest strategic challenge facing India today is systemic. India is trying to figure out its position in the contemporary international system, and because the system itself is in a state of flux, the complexities facing India are enormous. The debate about the nature of the post–Cold War international system has been going on for more than two decades now and still shows no signs of abating. Though scholars by and large accept that the United States is the dominant power in the world today, there are differences with regard to how far ahead the United States is relative to the other states and how long this dominance will last. Also, there is some question as to whether the United States is clearly ahead in all dimensions of power. There is an emerging consensus that a multipolar system is evolving in the Asia-Pacific where the center of gravity of global politics and economics is situated today.[14] Though the United States remains the predominant power in the region, its primacy is increasingly being challenged by China, making the region highly susceptible to future instability. China's future conduct is the great regional uncertainty even as it is also the most important factor affecting regional security.[15]

It is in this broader global and regional strategic context that India is trying to fashion its foreign policy. Throughout the Cold War, India jealously guarded its nonaligned foreign policy posture. After the fall of the Berlin Wall, the policy of nonalignment started to unravel because the two blocs that India wanted to guard its strategic autonomy against no longer existed. Today India is confronted with the challenge of redefining nonalignment. While rhetorically it may still make sense for India to proclaim its nonaligned status,[16] in practice it has no option but to cultivate its ties with major powers in the international system. The most controversial in the context of Indian domestic political landscape is India's growing closeness to the United States. While some in India are suggesting that India is on the verge of becoming a client state of the United States, India has been very careful to cultivate other major powers as well. The changes in the structure of the international system have also enabled India to pursue a "multivector" foreign and security policy, allowing it to strengthen its ties with all major global power centers, including the EU, Russia, China, and Japan. But the search for India's rightful place in the global balance of power continues, because India cannot continue with its multidimensional foreign policy for long without incurring significant costs. In particular, India's ties with a rapidly rising China are becoming ever more complicated, redefining Indian foreign policy priorities.

The China Challenge

With the world riveted by Chinese aggressiveness against Japan and Southeast Asian states in recent years, one country has not been surprised: India. After all, New Delhi has been grappling with the challenge of China's rapid rise for some time now. Bilateral ties between China and India nosedived so dramatically in 2009 that Indian strategists were even predicting "the year of the Chinese attack on India"; it was suggested that China would attack India by 2012 primarily to divert attention from its growing domestic troubles.[17] This suggestion received widespread coverage in the Indian media, which was more interested in sensationalizing the issue than investigating the claims.[18]

Meanwhile, the official Chinese media picked up the story and gave it another spin. It argued that though a Chinese attack on India is highly unlikely, a conflict between the two neighbors could occur in one scenario: an aggressive Indian policy toward China about their border dispute, forcing China to take military action.[19] The Chinese media went on to speculate that the "China will attack India" line might just be a pretext for India to deploy more troops to the border areas.[20]

This curious exchange reflects an uneasiness that exists between the two Asian giants as they continue their ascent in the global interstate hierarchy. Even as they sign loftily worded documents year after year, the distrust between the two is actually growing at an alarming rate.[21] Economic cooperation and bilateral political as well as sociocultural exchanges are at an all time high; China is India's largest trading partner. Yet this cooperation has done little to assuage each country's concerns about the other's intentions. The two sides are locked in a classic security dilemma, where any action taken by one is immediately interpreted by the other as a threat to its interests.

At the global level, the rhetoric is all about cooperation, and indeed the two sides have worked together on climate change, global trade negotiations, and demanding a restructuring of global financial institutions in view of the global economy's shifting center of gravity. At the bilateral level, however, mounting tensions reached an impasse when China took its territorial dispute with India all the way to the Asian Development Bank in 2009. There China blocked India's application for a loan that included money for development projects in the Indian state of Arunachal Pradesh, which China continues to claim as part of its own territory. Also, the suggestion by the Chinese to the U.S. Pacific Fleet commander in 2009 that the Indian Ocean should be recognized as a Chinese sphere of influence has raised hackles in New Delhi.[22] China's lack of support for the U.S.–India civilian nuclear energy cooperation pact, which it tried to block at the Nuclear Suppliers Group (NSG), and its obstructionist stance about bringing the terror masterminds of the November 2008 Mumbai attacks to justice have further strained ties.

Sino-Indian frictions are growing, and the potential for conflict remains high. Alarm is rising in India because of frequent and strident Chinese claims about the Line of Actual Control in Arunachal Pradesh, which China refers to as Southern Tibet, and Sikkim. Indians have complained of a dramatic rise in Chinese intrusions into Indian territory over the last few years, mostly along the border in Arunachal Pradesh where China has upped the ante on the border issue. It has been regularly protesting against the Indian prime minister's visit to Arunachal Pradesh, asserting its claims over the territory.[23] What has caught most observers of Sino-Indian ties by surprise, however, is the vehemence with which Beijing has contested recent Indian administrative and political action in the state, even denying visas to Indian citizens of Arunachal Pradesh.

The recent rounds of boundary negotiations have been a disappointing failure, with a growing perception in India that China is less willing to adhere to earlier political understandings about how to address the boundary dispute. Even the rhetoric has degenerated to such an extent that a Chinese analyst connected to China's Ministry of National Defense claimed in an article in 2009 that China could "dismember the so-called 'Indian Union' with one little move" into as many as thirty states.[24]

Pakistan, of course, has always been a crucial foreign policy asset for China, but with India's rise and U.S.–India rapprochement, its role in China's grand strategy is bound to grow even further. Not surprisingly, recent revelations about China's shift away from a three-decades' old cautious approach on Jammu and Kashmir, its increasing military presence in Pakistan, planning infrastructure linking Xinjiang and Gwadar, issuing stapled visas to residents of Jammu and Kashmir, and supplying nuclear reactors to Pakistan all confirm a new intensity behind China's old strategy of using Pakistan to secure its interests in the region.[25]

India's challenge remains formidable. Though it has not yet achieved the economic and political profile that China enjoys regionally and globally, India is increasingly bracketed with China as a rising or emerging power—or even a global superpower. Indian elites who have been obsessed with Pakistan for more than sixty years suddenly have found a new object of fascination. India's main security concern now is not the increasingly decrepit state of Pakistan but an ever more assertive China, a shift that is widely viewed inside India as one that can facilitate better strategic planning.

India's defeat at China's hands in 1962 shaped the Indian elite's perceptions of China, and they are unlikely to alter them anytime soon. China is thus viewed by India as a growing, aggressive nationalistic power whose ambitions are likely to reshape the contours of the regional and global balance of power with deleterious consequences for Indian interests. China's recent hardening toward India could well be a function of its own internal vulnerabilities, but that is hardly a consolation to Indian policy-makers who have to respond to an Indian public that increasingly wants India to assert itself in the region and beyond. India is rather belatedly gearing up to respond with its own diplomatic and military overtures, setting the stage for a Sino-Indian strategic rivalry.[26] Both India and China have a vested interest in stabilizing their relationship by seeking out issues on which their interests converge, but pursuing mutually desirable interests does not inevitably produce satisfactory solutions to strategic problems. A troubled history coupled with the structural uncertainties engendered by their simultaneous rise is propelling the two Asian giants into a trajectory that they might find rather difficult to navigate in the coming years.[27] Sino-Indian ties have entered turbulent times, and they are likely to remain there for the foreseeable future.

Managing China's Rise

The end of the Cold War has also enabled India to pursue a mutually beneficial relationship with the United States,[28] which is one of the ways in which

India is trying to manage China's rapid ascent in global interstate hierarchy. As a consequence, India's incorporation into the global nuclear order, something that India has desired for long, has become a real possibility. The present structure of the international system brings to the fore a factor central to Indian foreign and security policy in contemporary times—the centrality of the United States. The predominant position that the United States enjoys in the global hierarchy, notwithstanding all the debate about multipolarity, makes it central to Indian diplomacy. India has realized that it cannot find a place in the global nuclear order without the support of the United States. The U.S.–India civilian nuclear energy deal is as much about a global strategic realignment as it is about India's quest for energy security.[29] The United States remains vital to Indian interests in the near term, but India has yet to come to terms with all that it entails in being a "strategic partner" of the United States. Nowhere is this more evident than in the Indian policy toward the Middle East, where the United States has encouraged India and Israel to come closer but is strongly resisting India's burgeoning ties with Iran. The United States has also encouraged a larger role for India in East Asia, and India has actively tried to raise its profile in the region.

It was the end of the Cold War that really brought East Asia back to the forefront of India's foreign policy horizons. The disintegration of the Soviet Union radically transformed the structure of the then prevailing international system and brought to the fore new challenges and opportunities for countries such as India. India was forced to reorient its approach toward international affairs in general and toward East Asia in particular. The government of P. V. Narasimha Rao launched its Look East policy in the early 1990s explicitly to initiate Delhi's reengagement with East Asia.

Indian engagement of East Asia in the post–Cold War era has assumed significant proportions and remains a top foreign policy priority for the Indian policy-makers. India is now a full dialogue partner of the Association of South East Asian Nations (ASEAN) since 1995, a member of the ASEAN Regional Forum, the regional security forum since 1996, and is a founder member of the East Asian Summit launched in December 2005. India is also a summit partner of ASEAN on par with China, Japan, and South Korea since 2002. Over the years, India has also come to have extensive economic and trade linkages with various countries in the region, even as there has also been a gradual strengthening of security ties. The present Indian Prime Minister, Manmohan Singh, has made it clear that his government's foreign-policy priority will continue to be East and Southeast Asia, which are poised for sustained growth in the 21st century.

India's efforts to make itself relevant to the region come at a time of great turmoil in the Asian strategic landscape. Events over the last few years have underlined China's aggressive stance against rivals and U.S. allies in Asia, and there may be more tension to come. With its political and economic rise, Beijing has started trying to dictate the boundaries of acceptable behavior to its neighbors. As a result, regional states have already started reassessing their strategies, and a loose anti-China balancing coalition is emerging. India's role becomes critical in such an evolving balance of power. As Singapore's elder statesman Lee Kuan Yew has argued, he would like India to be "part of the Southeast Asia balance of forces" and "a counterweight [to China] in the Indian Ocean."[30]

Both New Delhi and Tokyo have made an effort in recent years to put Indo-Japanese ties into high gear. India's booming economy makes it an attractive trading and business partner for Japan as the latter tries to overcome its long years of economic stagnation. Japan is also reassessing its role as a security-provider in the region and beyond, and of all its neighbors, India seems most willing to acknowledge Japan's centrality in shaping the evolving Asia-Pacific security architecture. Moreover, a new generation of political leaders in India and Japan view each other with fresh eyes, allowing for a break from past policies that is changing the trajectory of bilateral relations.[31] While China's rise figures into the evolution of Indo-Japanese ties, so, too, does the U.S. attempt to build India into a major balancer in the region.

Both India and Japan are well aware of China's not so subtle attempts at preventing their rise. It is most clearly reflected in China's opposition to the expansion of the UN Security Council (UNSC) to include India and Japan as permanent members. China's status as a permanent member of the UNSC and as a nuclear weapon state is something that it would be loathe to share with any other state in Asia. India's Look East policy of active engagement with the ASEAN and East Asia remains largely predicated upon Japanese support. India's participation in the East Asia Summit was facilitated by Japan, and the East Asia Community proposed by Japan to counter China's proposal of an East Asia Free Trade Area also includes India. While China has resisted the inclusion of India, Australia, and New Zealand in the ASEAN, Japan has strongly backed the entry of all three nations.

The massive structural changes taking place in the geopolitical balance of power in the Asia-Pacific region are driving India and Japan into a relationship that is much closer than many could have anticipated even a few years back. The rapid rise of China in Asia and beyond is the main pivot even as New Delhi is seeking to expand economic integration and interdependence with the region. India is also developing strong security linkages with the region and trying to actively promote and participate in regional and multilateral initiatives. New Delhi's ambitious policy in East and Southeast Asia is aimed at significantly increasing its regional profile. Smaller states in the region are now looking to India to act as a balancer in view of China's growing influence and America's anticipated retrenchment from the region in the near future, while larger states see it as an attractive engine for regional growth. It remains to be seen if India can indeed live up to its full potential, as well as to the region's expectations.

A Strategic Void

As Indian foreign and security policies emerge out of the confines of the Cold War ideological straightjacket, it is still not clear if in the absence of a national consensus on India's strategic challenges, India can construct an apparatus that is adequate to productively engage with the world. Resources alone do not make a major power. Its quality of government must be willing and able to transform the potentialities of national power into a political reality. Good governance, as underlined by Morgenthau, is an independent requirement of national power, and in the case of India, the government has not been able to effectively strike a balance between its material and human resources and its foreign policy priorities.[32]

A major reason why India has been unwilling and/or unable to make effective use of its resources in supporting national policy in the realm of foreign affairs is a perceptible lack of institutionalization of the nation's foreign policy making. At its very foundation, Indian democracy is sustained by a range of institutions from the more formal ones of the executive, legislative, and the judiciary to the less formal ones of the broader civil society. It is these institutions primarily that have allowed Indian democracy to flourish for more than sixty years in contrast to the failure of democracy in many other societies. However, in foreign policy, the lack of institutionalization has resulted in a failure to take the long view.

Indian political elites have often boasted that there is a consensus on major foreign and security policy issues facing the country. Aside from the fact that such a consensus has been more of a result of intellectual laziness and apathy than any real attempt to forge a coherent grant strategy that cuts across ideological barriers, this is most certainly an exaggeration. Until the early 1990s, the Congress Party's dominance over the Indian political landscape was almost complete, and there was no political organization of an equal capacity that could bring to bear its influence on foreign and security policy issues in the same measure. It was the rise of the Hindu nationalist Bharatiya Janata Party (BJP) since the late 1980s that gave India a significantly different voice on foreign policy. But more important, it is the changes in the international environment since early 1990s that have forced Indian policy-makers to challenge some of the assumptions underlying their approach to the outside world.

In the initial years after independence, Nehru reigned so supreme in the realm of foreign policy that professionalism suffered. K.P.S. Menon, who served as Secretary in the Ministry of External Affairs, wrote in his autobiography:

> A Foreign Office is essentially a custodian of precedents. We had no precedents to fall back upon, because India had no foreign policy of her own until she became independent. We did not even have a section for historical research until I created one . . . Our policy therefore *necessarily rested on the intuition of one man*, who was Foreign Minister, Jawaharlal Nehru. *Fortunately his intuition was based on knowledge. . . .* " (emphasis added, throughout).[33]

It is not surprising that after interviewing former members of the Indian Foreign Service, Waner F. Ilchman noted a "tendency for men in the field to write what the Prime Minister wished to hear."[34] Even after Nehru's death, foreign policy continues to remain the preserve of successive prime ministers, and no effort was made to construct an effective national security apparatus.

Some have blamed Nehru for his unwillingness to construct a strategic planning architecture because he single-handedly shaped Indian foreign policy during his tenure,[35] but even his successors have failed to pursue institutionalization in a consistent manner. The BJP-led National Democratic Alliance (NDA) came to power in 1999, promising that it would establish a National Security Council (NSC) to analyze the military, economic, and political threats to the nation and to advise the government on meeting these challenges effectively.

The party did set up the NSC in the late 1990s, defining its role in policy formulation, but it nonetheless failed to institutionalize the NSC or to provide it with the capabilities necessary to play the role assigned to it. As in the past, important national security decisions were addressed in an ad hoc manner without utilizing the Cabinet Committee on Security, the Strategic Policy Group (comprised of key secretaries, service chiefs, and heads of intelligence agencies), and officials of the National Security Advisory Board. Moreover, as has been rightly pointed out, the structure of the NSC makes long-term planning impossible, thereby negating the very purpose of its existence. Its effectiveness remains hostage to the weight of the National Security Advisor (NSA) in national politics.[36] The NSA has become the most powerful authority on national security, eclipsing the NSC as an institution.

When the Congress-led United Progressive Alliance came to power in 2004, it also promised to make the NSC a professional and effective institution (and blamed the NDA for making only cosmetic changes in the institutional arrangements), but to date it has failed to make it work. The NSC still does not anticipate national security threats, coordinate the management of national security, and engender long-term planning by generating new and bold ideas. An effective foreign policy institutional framework would not only identify the challenges but would develop a coherent strategy to deal with it, organize the bureaucracy, and persuade the public. The NSC by itself is not a panacea, particularly in light of the inability of the NSC in the United States to successfully mediate in the bureaucratic wars and effectively coordinate policy. But the lack of an effective NSC in India is merely a symptom of the continuing inability and unwillingness of India's policy-makers, across political ideologies, to provide institutional underpinnings to Indian foreign policy. The NDA government had constituted a Group of Ministers (GoM) in 2000 to review the national security system in its entirety. This GoM had recommended the formulation of a national security doctrine, Defence Minister's Directive, Long Term Integrated Perspective Plan (LTIPP) and the Joint Services Plan.[37] But these recommendations have not been operationalized. It is often assumed that India has the necessary institutional wherewithal to translate its growing economic and military capabilities into global influence even though the Indian state continues to suffer from weak administrative capacity in most areas of policymaking. This lack of institutionalization will continue to affect Indian behavior in the realm of foreign and security policy.

The institutional mechanisms are deficient, leading to the absence of an overall strategic framework for Indian national security policy. Moreover, the Indian political leadership is consumed by day-to-day, sometime seemingly intractable, domestic issues. But as India's profile and stature have risen in the international system, the fissures in foreign and security policy issues are out in the open. India is debating the choices it faces on foreign policy as it has never done before. In the absence of strategic clarity, the policy-makers have found it difficult to manage Indian bureaucracy, which has been resistant to change in the face of rapidly changing international context. This has resulted in a lack of coordination on national security issues, and a sense of drift prevails on a whole host of issues.

India's foreign policy elite remains mired in the exigencies of day-to-day pressures emanating from the immediate challenges at hand rather than evolving a grand strategy that integrates the nation's multiple policy strands into a cohesive whole to be able to preserve and enhance Indian interests in a rapidly changing global environment. The Indian elites do have a growing sense of their country as an emerging great power, as an important player on the global stage. Yet the Indian State seems unable to leverage the opportunities presented by India's economic rise to its full potential. Tensions are inherent between the requirements of a Great Power foreign policy and complications of a democratic multinational state. Policy-making in democracies is often a messy process, full of complexities. but there is a "near paralytic fragmentation of authority" in Indian polity to the point where on an entire range of crucial issues a sense of drift prevails.[38] Policy initiatives continue to be framed in a political environment that is highly fragmented and unstable.

Reflections

Kishore Mahbubani in his book *The New Asian Hemisphere* makes a strong case for India's global leadership, asserting that India's "credentials as the world's largest democracy; its open, tolerant and inclusive culture; its unique geopolitical and cultural position as a bridge between east and west gives it a unique opportunity to provide the leadership for forging new forms of global governance that spaceship Earth desperately needs as it sails into the future."[39] Certainly, India's rise to global prominence may not be very problematic for the world, as its democratic political system will go a long way in allaying the apprehensions of the established powers. The real challenge for India lies in the domestic sphere, where the Indian state will have to succeed in overcoming the constraints that continue to inhibit India's potential.

Examining India's approach to national security as a whole in the post–Cold War period, one finds there is a significant move away from the ideological moorings of the Cold War. From a sense of idealism that pervaded Indian foreign policy and national security policy-making then, there is now a greater sense of "strategic realism." But a lack of strategic orientation in foreign and security policy making often results in a paradoxical situation where, on the one hand, India is accused by various domestic constituencies of angering this or that country by its actions, while on the other hand, India's relationship with almost all major powers is termed as a "strategic partnership" by the Indian government. India is perhaps the only major global power that has never produced a national security strategy document. Ad hoc-ism remains the norm in national security decision making, and this will continue to circumscribe India's rise in the global order.

Notes

1. Chidanand Rajghatta and Prabhakar Sinha, "India Buys 200 Tons Gold from IMF," *Times of India*, November 4, 2009.

2. Ibid.

3. Rishi Shah, "India May Beat China Next Year: World Bank," *Economic Times*, October 22, 2010.

4. Harsh V. Pant, *Contemporary Debates in Indian Foreign and Security Policy: India Negotiates Its Rise in the International System* (New York: Palgrave Macmillan, 2008), pp. 1–15.

5. Devin Hagerty, "India and the Global Balance of Power: A Neorealist Snapshot," in Harsh V. Pant (ed.), *Indian Foreign Policy in a Unipolar World* (London: Routledge, 2009).

6. Michael Mandelbaum, *The Fates of Nations: The Search for National Security in the Nineteenth and Twentieth Centuries* (Cambridge: Cambridge University Press, 1988), pp. 2, 4.

7. Christopher Layne, "The Unipolar Illusion: Why New Great Powers Will Rise?," *International Security*, vol. 17, no. 4 (Spring 1993), pp. 9–10.

8. C. Raja Mohan, "India's Grand Strategy in the Gulf," India in the Gulf Project, The Nixon Center.

9. AICC File No. 8, 1927, Nehru Memorial Museum and Library.

10. Extracts from Nehru's speech to the Constituent Assembly of India, December 4, 1947, partially reproduced in A. Appadorai, *Select Documents on India's Foreign Policy and Relations 1947–1972, Vol. 1* (Oxford: Oxford University Press, 1982), p. 10.

11. Sisir Gupta, "Great Power Relations and the Third World," in Carsten Holbraad (ed.), *Super Powers and World Order* (Canberra: Australian National University Press, 1971).

12. K. Subrahmanyam, *Indian Security Perspectives* (New Delhi: ABC Publishing House, 1982), p. 127.

13. Ibid., p. 129.

14. See, for example, Fareed Zakaria, *The Post-American World* (New York: W.W. Norton & Co., 2008).

15. Harsh V. Pant, *China's Rising Global Profile: The Great Power Tradition* (Portland, Oreg.: Sussex Academic Press, 2011), pp. 11–28.

16. "India to U.S.: NAM Still Relevant," *Indian Express*, June 30, 2007.

17. "Nervous China May Attack India by 2012: Defence Expert," *Indian Express*, July 12, 2009.

18. See for example, http://www.ndtv.com/news/india/china_could_attack_india_by_2012_defence_journal.php

19. Li Hongmei, "Veiled Threat or Good Neighbour?" *People's Daily*, June 19, 2009.

20. Ibid.

21. For a detailed discussion of the present state of Sino-India relations, see Harsh V. Pant, *The China Syndrome: Grappling with an Uneasy Relationship* (New Delhi: HarperCollins, 2010).

22. Yuriko Koike, "The Struggle for Mastery of the Pacific," *Project Syndicate*, May 12, 2010.

23. "China 'Strongly' Protests over PM's Visit to Arunachal Pradesh," *Press Trust of India*, October 13, 2009.

24. James Lamont and Kathrin Hille, "Chinese Essay Sparks Outcry in India," *Financial Times*, August 12, 2009.

25. "China to Build Two Nuclear Reactors in Pakistan," *Hindustan Times*, April 29, 2010.

26. Harsh V. Pant, "India Comes to Terms with a Rising China," in Ashley J. Tellis et al., *Asia Responds to Its Rising Powers* (Washington, D.C.: The National Bureau of Asian Research, 2011), pp. 101–128.

27. For a detailed examination of the various dimensions of China's rise and their implications for India, see Harsh V. Pant (ed.), *The Rise of China: Implications for India* (New Delhi: Cambridge University Press, 2012).

28. S. Paul Kapur and Sumit Ganguly, "The Transformation of U.S.–India Relations: An Explanation for the Rapprochement and Prospects for the Future," *Asian Survey*, vol. 47, no. 4, pp. 642–656.

29. Harsh V. Pant, *The US-India Nuclear Pact: Policy, Process and Great Power Politics* (Oxford: Oxford University Press, 2011).

30. P. S. Suryanarayana, "China and India Cannot Go to War: Lee Kuan Yew," *The Hindu*, January 24, 2011.

31. On the rapidly changing trajectory of India-Japan ties in recent years, see Harsh V. Pant, "India and Japan: A Slow, But Steady, Transformation," in Sumit Ganguly (ed.), *Indian Foreign Policy: Retrospect and Prospect* (New Delhi: Oxford University Press, 2010), pp. 206–225.

32. Hans J. Morgenthau, *Politics Among Nations: The Struggle for Power and Peace* (New York: Alfred A. Knopf, 1962), pp. 143–148.

33. K.P.S Menon, *Many Worlds* (New Delhi: Oxford University Press, 1965), p. 271.

34. Warren F. Ilchman, *Journal of Commonwealth Political Studies*, November 1966 (Leicester University Press).

35. Singh, *Defending India*, p. 34.

36. Ashley J. Tellis, *India's Emerging Nuclear Posture: Between Recessed Deterrent and Ready Arsenal* (New York: Oxford University Press, 2001), p. 658.

37. Group of Ministers Report, *Reforming the National Security System: Report of the Group of Ministers in National Security* (New Delhi: Government of India, 2001), pp. 98 and 107–108.

38. Pratap Bhanu Mehta, "Not So Credible India," *Indian Express*, April 24, 2008.

39. Kishore Mahbubani, *The New Asian Hemisphere: The Irresistible Shift of Global Power to the East* (New York: Public Affairs, 2008).

11 Russia: A Fallen Superpower Struggles Back

Robert H. Donaldson

THE RUSSIAN FEDERATION HAS EXISTED IN ITS PRESENT FORM AND boundaries for only two decades. In such a short time Russia has taken impressive steps toward remaking its economic and political systems and in defining a new role in the post–Cold War world. But not surprisingly, there is much still undone and undecided about the requirements for Russia's security.

This chapter will explore the perceptions of Russian leaders regarding their country's position and role in the global and regional security systems, examining both the military and economic dimensions of security and both internal and external threats. It will consider not only how these perceptions and Russia's actual capabilities have changed over time, but also how the Russian state has been organized to conduct its security policies.

Adjusting to a Changing International System

Partly in response to major shifts in the international environment, Soviet President Mikhail Gorbachev made fundamental changes in Moscow's foreign policy before the collapse of the USSR at the end of 1991. The tight bipolarity of the late 1940s had given way to a looser system. Conflicts between the Soviet Union and other communist states, beginning in Yugoslavia in 1948 and climaxing in China in the late 1960s, had fractured the once-monolithic communist subsystem. Thermonuclear weapons and ICBMs had raised the costs of actual warfare between the chief adversaries to unacceptable levels.[1]

Having achieved a state of strategic military parity with the United States in the 1970s, the USSR imposed monumental stresses on its economy in subsequent years in its efforts to keep up in the arms race. Indeed, because the USSR was largely isolated from the global economic system, its very status as a superpower was called into question in the late 20th century, as military strength became less relevant to international issues than economic capability.[2]

Russian Federation President Boris Yeltsin inherited these features of a changed international system, as well as territorial boundaries markedly different not only from those of the USSR but also from those of any prior independent Russian state. Lands near the Baltic Sea and the Black Sea, acquired by Peter I and Catherine II through decades of warfare, now belonged to units of the former USSR that were no longer under Moscow's control. Except for the province of Kaliningrad, separated from the rest of Russia by independent Lithuania, the Russian Federation had no border with the East European states that had been part of Stalin's postwar empire. Not only was Russia's frontier more distant from the capitals of Europe, but a Russia separated from the now-independent republics of the Caucasus and Central Asia was also no longer bordered by the states of the Middle East and South Asia.

The United States had emerged from the Cold War as the globe's only surviving superpower, but initially the leaders of the new Russia perceived no threat from it. Indeed, Russia's rulers faced a less-threatening environment beyond the borders of the former empire than had been the case for centuries. Far more vexing were Russia's newly independent neighbors, the former Soviet republics. For the ministries and personnel of the new Russian state were largely inherited from the Soviet Union, and it was the domestic and not the foreign policy bureaucracies that had the mechanisms and expertise to deal with the former Soviet republics. Moreover, these new neighbors posed a pressing challenge, since they encompassed many areas of instability and ethnic conflict.

The abandonment of Marxist-Leninist ideology, combined with the fall of the Soviet Communist Party and the disintegration of the USSR itself in 1991, left a conceptual void in the foreign policy of the newly independent Russian Federation that raised to the forefront the question of Russia's national identity. Russia had never existed as a nation-state; rather, during both the tsarist and the Soviet periods it had been a multinational empire with messianic ambitions. Moreover, the tsars, unlike the rulers of Britain or France, had colonized lands that bordered on their home territories, thus producing an unusual admixture of Russian and non-Russian peoples.[3]

Now the proportion of ethnic Russians in the total population of the Russian Federation exceeds 80 percent, whereas they comprised just over half the population of the USSR in its last years. But the non-Slavic minorities of Russia are the fastest growing and most restive part of a declining population. Further complicating the definition of Russia's national identity is the fact that, at the time the USSR disappeared, 25 million ethnic Russians lived outside the Russian Federation, in the other newly independent states of the former Soviet Union.

Defining a New National Purpose

Not only were the people of the Russian Federation experiencing new geopolitical confines, but they were also painfully aware of the relative weakness of their state, in comparison to the superpower status enjoyed by the USSR at the height of its power. The precipitous economic decline of the early 1990s produced a profound sense of national humiliation, as Russia's leaders—first Gorbachev, then Yeltsin—appeared to be begging Western leaders for aid. This combination of a loss of national mission, a wounded national pride, and a confused national identity has underscored the absence of a sense of national purpose in the foreign policy of the new Russia.[4]

Despite the elimination of an official ideology, the tradition of expressing the basic principles of policy in an official programmatic statement has persisted in post-Soviet Russia. There were lengthy delays and vigorous debates before Boris Yeltsin could approve "Concept" documents on Russian foreign policy (1993), military doctrine (1993), and national security (1997). Vladimir Putin approved new versions of each of these within the first six months of 2000.[5] During the remaining years of Putin's presidency, he chose to amplify and update policy guidance by means of authoritative speeches presented before audiences of military leaders or ambassadors.[6]

A return to the issuance of official "concept" documents came early in Dmitri Medvedev's presidency, with the promulgation of a new Foreign Policy Concept in 2008[7] and his approval in May 2009 of a *National Security Strategy of the Russian Federation to 2020* (hereafter, NSS).[8] This latter document differs markedly both in tone and content from the 1999 version, reflecting the greater confidence in Moscow that resulted in large part from the economic and political stabilization achieved in the intervening decade, aided in no small way by the spiraling prices Russia received for its considerable oil and gas exports.

Indeed, Vladimir Putin had argued, in a dissertation he submitted in the mid-1990s for a graduate degree, that energy resources and other raw material reserves could serve as the lever for projecting Russia back to a position of international influence,[9] and this belief is plainly articulated in the NSS: "In the long term, the attention of international politics will be focused on ownership of energy resources, including those in the Near East, the Barents Sea shelf and other parts of the Arctic, in the Caspian basin, and in Central Asia."[10] And though the country was clearly no longer a superpower, it was increasingly capable of acting as a Great Power—still possessed of "strategic sufficiency," dominant in its own region and participating diplomatically and economically on a global basis, enabled by its membership in key multilateral groupings with global reach, especially the UNSC, the G-8, and BRICs (a loose grouping of rising powers, comprising Brazil, Russia, India and China).

In short, Russia aspires to be a key multiregional player in a multipolar world, as the NSS clearly states: "The transition in the international system from opposing blocs to principles of multivector diplomacy, together with Russia's resource potential and pragmatic policy for its use, have broadened the possibilities for the Russian Federation to reinforce its influence on the world stage."[11]

The Realms of Russian Security Policy

In the realm of economics, the NSS expresses an aspiration for Russia to move into the ranks of the top five economies globally. There is, however, a downside to this goal: Russia is now sufficiently integrated into the world economy that it is subject to being buffeted by adverse financial and economic developments beyond its borders. Russia's GDP plunged during the 2008–2009 global financial crisis, harmed not only by the drop in energy prices but also by the weakness of its manufacturing sector. Russian leaders (Medvedev's advisors more vocally than Putin's) have recognized the vulnerability inherent in an economy that is primarily a raw materials exporter, and they seek investments and innovations that would enable a more balanced development ("modernization"), especially in "new" technologies.

In the realm of trade, Russia was handicapped by being the last major economic entity to become a member of the Word Trade Organization, finally joining in 2012. Major issues included Russia's policy of maintaining dual pricing between domestic users and exports of natural gas, its lack of protection of trade-related intellectual property rights, and restrictions on foreign direct investment in Russia's services sectors.

Militarily, the first Chechen War (1994–1996) dramatically exposed the fact that the country's demoralized armed forces had not recovered from their defeat in Afghanistan and were still not capable of sustained offensive operations. In the absence of thoroughgoing reform, the state of Russia's military at the end of the 1990s was alarming. As one American expert described it:

> Discipline has collapsed, equipment is becoming antiquated, morale has sunk to an all-time low, good officers and non-commissioned officers are leaving the service, the country's generals have been politicized, and Moscow's ability to ensure the military's obedience in a crisis is doubtful.[12]

Russian forces, employing overwhelming force and brutal tactics, were able to impose a precarious peace (under a Kremlin-backed warlord) in Chechnya as a result of a second war, but again their performance was hardly impressive. Even its short war with Georgia in August 2008, ostensibly a military triumph for Russia, in fact revealed substantial deficiencies in its military operations. We will return to this subject in the next section, when we discuss Russia's official military doctrine.

Russia and the West

Being acutely aware of their country's descent from the status of multinational empire existing as one of two superpowers during the era of the Cold War, the Russian leadership has been particularly sensitive to what it has perceived as a unilateralism on the part of the "sole remaining superpower." Russia's official security policy documents since the mid-1990s have been dominated by the theme of the inadmissibility of unilateralism and the necessity of a multipolar world. In this context, it has insisted on the need to deliberate on and decide conflict situations in the UNSC, where it exercises the veto power of the former USSR.

Russia has perceived NATO as the instrument of choice of U.S. officials, who promote a unilateralist strategy. When, as a result of domestic politics in the United States and bureaucratic momentum in the alliance, Poland, the

Czech Republic, and Hungary were accepted into NATO in 1997, there was nothing that Yeltsin could do about it except complain loudly and demand a halt to further expansion. The offensive war waged by NATO against Serbia over Kosovo in 1999 was a traumatic event for Moscow, which loudly objected to the assault on the sovereign rights of its traditional ally as well as to the bypassing of the UN. Moscow complained bitterly about the possible adverse precedent for dealing with various "frozen conflicts" in the former Soviet space; these complaints became more acute as Kosovo moved closer to a declaration of independence.[13]

During a brief period early in his presidency, Putin hinted that Russia might actually welcome an invitation to join NATO.[14] But he quickly backed away, and the NATO-Russia Council was set up in 2002 as a forum for cooperation. When the second round of enlargement took place in 2004, including the three Baltic republics, Russia again objected. But it was only when enlargement advocates began pushing for the entry of Georgia and Ukraine that the temperature again reached the boiling point. The proposed intrusion of the U.S.-led military alliance into a very sensitive area viewed by Russia as vital to its security impelled Russia's new President Medvedev in 2008 to invite discussions on an alternative security system, extending from Vancouver to Vladivostok, which would embody the principle that "no one should strengthen their security at the expense of the security of others." Russia has rightly seen itself as excluded from the European security "architecture," and it seeks guarantees (if not a veto power) to ensure its participation (and to ensure that its perceived security interests are not threatened) in European security.[15]

It was not until early 2009 that any cooperative relationship between NATO and Russia resumed. Nevertheless, the August 2008 Russo-Georgian war (and subsequent reports that placed the blame for its initiation on the Georgian government) alarmed some of NATO's European members to the point that further enlargement has been put on indefinite hold.

Anne Clunan has persuasively argued that Russia's attitudes toward NATO can best be explained not from a "realist" assessment of a security threat, but rather from what she calls "aspirational constructivism." It was not until NATO began to implement a plan for enlargement that Russia raised objections, since it perceived enlargement as "threats to Russia's rightful place in European and global politics. . . . " Russia's elites perceived that threat "not primarily in military terms but in status terms . . . Russian policies on

European security became reactive to perceived Western efforts to undermine Russia's aspirations regarding its global status." According to this interpretation, Russia is not a "revisionist power seeking to challenge" the United States and the West but instead it "seeks to join the West, but in a manner that allows its leaders to maintain national self-esteem in the eyes of Russian political elites, primarily through Russia's involvement in the management of global affairs and its partial Westernization."[16]

Russia's attempt in both 1994 and 2000 to open a discussion about gaining an "equal partnership" or other form of privileged position with NATO is confirmation of Clunan's assertion. Martin Smith's book-length study on the issue also supports the notion that Russia's stance has not been inherently anti-Western. Rather, as he puts it, the objective under both Yeltsin and Putin "appears more to have been to secure an institutionalized cooperative status and hence influence—for Russia *vis-à-vis* the West."[17] A very similar explanation of the Russian position was offered by Sergei Karaganov and Timofei Bordachev:

> Russia's elite, which views itself as the victor in the struggle against totalitarian Communism, has never considered its country defeated in the Cold War. Meanwhile, the West has been trying to treat it as a defeated country, which has laid a deep foundation for a new and potentially rough confrontation. The solution is simple: the Old West will have to either try to 'finish off' Russia or to conclude an honorable peace with it and thus finish, once and for all, the 'unfinished' Cold War.[18]

Throughout the past two decades, Russia has demonstrated awareness that the "West" itself is not monolithic, and Moscow has occasionally reverted to the divide and conquer tactic often utilized by Soviet leaders. Yeltsin sought to build a special relationship with France and Germany, based on a presumed distinction between Europe's interests and those of the United States, even on occasion stressing that "we" Europeans "do not need an uncle from somewhere" to lead on our behalf.[19] Putin followed this same path, especially with Gerhard Schroeder's Germany, capitalizing on his own KGB experience in Germany and his knowledge of the language. Most notably, Chirac's France, Schroeder's Germany, and Putin's Russia combined to oppose the U.S. invasion of Iraq in 2003. President Sarkozy and Chancellor Merkel were less receptive to such anti-Washington groupings. And while the Russians would prefer to deal with European states on a bilateral basis, the EU as a whole has recently taken a more active role in resisting Russian efforts to gain a privileged economic position, particularly in the sensitive arena of energy.

Russia and Eurasia

In the former Soviet space, understandably its primary security concern, Russia has been on the defensive since the dissolution of the USSR. The CIS has proved to be a hollow shell, largely because the other members rightly feared Russia's domination of the organization. Accordingly, Russia has organized subgroupings in both economic and security spheres. The Eurasian Economic Community (EurAsEC) is a customs union, created in 2000, joining Russia with Belarus, Kazakhstan, Kyrgyzstan, and Tajikistan. The Collective Security Treaty Organization (CSTO) is a military alliance of these same states plus Armenia. From early discussions of a Monroe Doctrine in Russia's "near abroad"[20] and Yeltsin's efforts to have Russia appointed "peacekeeper" in the FSU,[21] to Medvedev's later claims for a "special" security sphere (a "zone of privileged interests"),[22] Russia has demonstrated its determination not to lose predominant influence in this region, where not only are its economic interests at stake but ethnic conflicts and secessionist movements are seen as threats (and opportunities for "outside" meddling). This was the essential meaning of the 2008 crisis with Georgia. But Russia has angrily resisted both NATO's threatened incursions and the competition for influence, especially in the oil-rich areas of the Caucasus and Central Asia, from the United States (and, *sotto voce*, from China). Russian policy in this key region is summed up in the NSS:

> The development of bilateral and multilateral cooperation with member states of the CIS is a priority direction of Russian foreign policy. Russia will seek to develop the potential for regional and sub-regional integration and coordination among member-states of the CIS, first of all within the framework of the Commonwealth, and also the CSTO and EurAsEC which exert a stabilizing influence on the overall situation in the regions bordering on the CIS . . . Of particular significance for Russia will be the reinforcement of the political potential of the SCO [Shanghai Cooperation Organization, joining China, Russia, and four of the Central Asian republics, discussed below] and the stimulation within its framework of practical steps toward the enhancement of mutual trust and partnership in the Central Asian region.[23]

Russia in Asia

Most of Russia's territory is in Asia, but Moscow's influence there seems to have declined significantly. Russia remains important as a source of energy for East Asia, as the major seller of arms for China and India, and as a key participant in security organizations with Asian states. But population decline, a cumbersome military, a stagnant economy, and a foreign policy focus increasingly turned westward raise serious doubts about whether Russia continues to play a significant role as a power in Asia.

In the latter half of the 1990s, concern over NATO's enlargement led foreign minister Yevgeny Primakov to champion an Asia-centered rival coalition of Russia, China, and India.[24] Yeltsin had negotiated settlement of Russia's border disputes with China (a process completed under Putin), and he had reiterated Moscow's desire to continue its close Cold War–era association with India. But a size-diminished and economy-weakened Russia had little to offer China and India in the 1990s beyond exports of advanced weaponry—for which the two Asian powers were Moscow's chief customers. Moreover, the border conflict between Beijing and New Delhi, which had produced war in 1962, had never been resolved, and even though relations between the two Asian giants have been much improved in recent years, it is still extremely difficult to envision an alliance between them (especially one aimed at balancing the United States).

With East Asia's other great power, Japan, the USSR had tenuous relations, due in part to Japan's U.S. security ties but also because territorial disputes left over from the closing days of World War 2 have never been resolved, leaving Tokyo and Moscow without a formal peace treaty. Just prior to the fall of the USSR, Yeltsin had hinted that a compromise on the disputed Kuril Islands could be arranged if he came to power. But the high hopes in Tokyo were soon dashed, as Yeltsin faced strong opposition from his parliament for any territorial concessions. Throughout the 1990s successive summits failed to bring progress, despite the fact that Russia desperately needed the investment in its Far East that the Japanese were dangling as a carrot.[25] In 2010, President Medvedev journeyed to the Kuril Islands (the first time a Russian ruler had done so), much to the displeasure of Japan, to reiterate Russia's determination to hold on to the territory.[26]

Gorbachev and Yeltsin fashioned a virtual "reversal of alliances" with the two states on the Korean peninsula. Moscow walked away from its security treaty assurances to North Korea while forging a rapprochement with the South, again in hopes of substantial economic investment. Yeltsin's first trip to the Far East as Russian president was to Seoul, not Pyongyang, Beijing, or Tokyo. Putin was more attentive to North Korea than his predecessors, but his efforts were devoted not toward renewal of defense commitments but toward achieving a breakthrough in multilateral talks aimed at ending sanctions against Pyongyang in return for its abandonment of its nuclear weapons program.

Part of the problem, of course, is that Russia's profile on the continent has been vastly overshadowed by the stunning economic rise of Asia's two most populous countries: China and India. Ironically, the rising military capabilities of China and India have been greatly aided by Russia, the country they now both so overshadow in Asia. Between them, the two have bought about 70 percent of post-Soviet Russia's arms exports. Chinese and Indian imports have declined as their domestic arms manufacturing capabilities have grown, aided by Russian-licensed technology. Neither is likely to wean itself entirely from Russian arms exports, but Russia's relative importance as a supplier has undoubtedly waned.

The main attraction of the Russian economy for China (as for much of the rest of Asia) is no longer arms but energy. Russia's vast oil and gas reserves are an irresistible alternative to Middle East supplies for both China and Japan, and oil now accounts for more than half of Russia's exports to Asia. Russia's new oil pipeline from Western Siberian fields to the Pacific, with a spur to China, made it possible for one-fourth of Russia's oil exports to go to Asia by 2012. China and Japan have been rivals in their hunger for Russian oil, and both have made investments to help spur Russian energy exploration and transport.[27] China is also importing large amounts of electricity from the Russian Far East.

Whereas Russia has become ever more important in Asia as a raw materials exporter, it figures hardly at all in the security picture. In great contrast to the USSR's position in Asia during the Cold War, one can find hardly any reference at all to Russia in discussions of the military balance in Asia in the 21st century.[28] Part of this is a result of Russia's own greater insularity. With limited resources and a need to give attention to domestic economic issues, Russia's leaders tend to focus foreign policy energies on the countries that are the nearest and most immediate threats to their security. In Asia, this results in greater attention to Central Asia and less to East, Southeast, and South Asia.

From Moscow's standpoint, the SCO, comprised of Russia, China, and all of the Central Asian states except Turkmenistan, is especially relevant to its security needs. This organization has focused on threats to the internal stability of its members, and particularly on the threat posed by Islamic extremism. It is undergirded by the 2001 treaty of "good neighborliness and cooperation" between Moscow and Beijing, which recently produced the first joint military exercises between the two countries in decades.[29]

Moscow's own CSTO, to which its closest allies among the former Soviet republics belong, including several in Asia, seemed for a while to be the preferred "balancer" to an expanded NATO, but the organization demonstrated its toothlessness in 2010 when it proved incapable of providing any stabilizing response to the serious crisis in one of the member states, Kyrgyzstan. Indeed, the improvement in that same year in relations between Russia and NATO, in the wake of the hiatus in NATO enlargement, has diminished the potential importance of the CSTO as a "rival" (if certainly not a "balancing") coalition to NATO.

So much of what is seen as Russia's diminished role in Asia is a function of China's rise and of the assumption that this comes in the form of a zero-sum game: China's increased power is at the expense of Russia's power in Asia. In 2012, this issue was put to Vladimir Putin by foreign experts who were meeting him in the annual Valdai Club discussions. Putin's reply is revealing:

> Foreign experts keep telling us about the threat from China. We are not worried at all. There is the huge Far East, Eastern Siberia, an under-populated territory. And there is powerful China, over a billion people. We should be afraid. We are not afraid. There is no threat on the side of China . . . We have co-existed with China for a thousand years. We had difficult moments, and at times better relations, but we know each other very well and we have got used to respecting each other. China does not have to populate the Far East and eastern Siberia to get what it needs: natural resources. We have just finished the construction of an oil pipeline. We are ready to build two gas pipelines. We will be supplying coal to them . . . China does not want to worsen relations with us to solve its current goals.[30]

Was this foreign policy realism, or was it bravado? The very fact that Putin felt compelled to reply to the question was perhaps indicative of the pervasive uncertainty that surrounds Russia's current position in Asia.

Russia and the Middle East

The southward expansion of the Russian Empire in the 18th and 19th centuries brought it into frequent conflict with the Ottoman Empire and with Persia, as well as with the competing empires of Britain and France. During the Soviet period, Lenin, Stalin, and their successors engaged in active (and occasionally expansionist) diplomacy in the region, as Moscow competed for territory and influence with the countries of the Middle East and with Western "imperialism." Although, after the collapse of the USSR, the Russian Federation now finds itself geographically separated from the Middle East by the buffer states of the Caucasus and Central Asia, Moscow's vital interests in these former Soviet republics and the vast energy reserves they contain—second only to those in the Middle East itself—have made the region a high priority for Russian foreign policy.

Accordingly, of highest concern in the region are Turkey and Iran, the two states that share a border with the states of the former Soviet Union and that are thus seen as Russia's potential competitors there. Russia's policies toward them have varied, in part as a result of the shifting political winds in Moscow, with Westernizers and nationalists competing over the direction of foreign policy. There has also been variability in Moscow's perception of the threat that Ankara or Tehran is seeking to extend pan-Turkic or Islamic-fundamentalist influence in the Caucasus and Central Asia at Russia's expense. In both cases, Russian policy has swung back and forth from nervous rivalry to a relationship so cooperative that it has included the sale of arms.[31]

Differing estimates of Iran's nuclear ambitions have generated considerable tension between Russia and the West. After lengthy delays, Russia (and China), in response to allegations by the International Atomic Energy Agency that Tehran was concealing some of its nuclear activities, finally withdrew their opposition in the UN Security Council to the imposition of sanctions. Though Moscow wants to persuade Iran not to acquire nuclear weapons, Russia is a beneficiary of trade with Iran, and it has not been willing to tighten the economic noose on Tehran as tightly as Washington would wish.

The turbulence in the Middle East during the Arab Spring uprisings and its aftermath again exposed the sharp differences between Moscow's perspectives and those of the West. Russia was persuaded to abstain on a UN resolution authorizing member states to take "all necessary measures" to protect Libyan civilians threatened by Qaddafi's forces. But Moscow felt that it had been deceived when air strikes by NATO and some Arab states were extended to play an active role in the overthrow of Qaddafi's regime. Russia strongly disagrees with the proposition that the "responsibility to protect" doctrine permits the international community to undertake regime change in a sovereign state. Accordingly, Moscow stood firm against UN draft resolutions that would allow foreign intervention against Russia's sole remaining Arab ally, the regime of Bashar Assad in Syria. Buttressing this stand for principle, as Syria's civil war became ever more bloody, was Russia's self-interest, given its large economic and military stake in Syria.

Russia's National Security Apparatus

The constitution of the Russian Federation, adopted by referendum in December 1993, clearly establishes the president as the chief decision-maker on matters of national security. The president's direct role in foreign policy–making is further enhanced by the fact that the foreign minister and the "power ministers" of Russia—the ministers of defense and interior, and the heads of the intelligence and security services—report to the president directly rather than through the prime minister. With so many responsible officials having direct access to the president—and given that Yeltsin's style was to keep his hands directly in so many matters—it is not surprising that there was an endemic messiness in Russian foreign policy during the Yeltsin years. Vladimir Putin, with the more disciplined style of a former KGB officer, brought a significant dose of much-needed orderliness to the Kremlin.

The creation of the Security Council in April 1992 was intended to bring the top foreign policy and national security officials together to deliberate and prepare decisions for the president to implement by decree. Twelve men (several of them prominent political figures) held the position of secretary of the Security Council during its first dozen years of existence. The many twists and turns in the short history of the Security Council make clear that some of its more politically ambitious secretaries have sought to inflate its authority, turning it into a powerful decision-making body with broadly defined operational responsibilities.[32] Boris Yeltsin clearly resisted these efforts, evidently sensing that such a body—reminiscent of the post-Stalinist Politburo—would dilute his own power. Vladimir Putin, a former occupant of the position, followed the pattern of enlarging the council's role only at times that he fully trusted its leadership.

Nevertheless, the Security Council remains significant if only because of the important positions that are represented on the council. As enumerated in a presidential decree of April 2004, these were the president, as chairman; the secretary; the prime minister; speakers of the two houses of the Federal Assembly; the foreign, defense, and interior ministers; the heads of both the domestic and foreign intelligence services; the chief of staff of the presidential administration; the head of the Russian Academy of Sciences; and the seven presidential representatives to the "super-regions" of Russia. Much of the work of the Security Council is conducted by interagency commissions, on which each relevant ministry is represented by a deputy minister.

Much of the focus of the Security Council has been on internal challenges. Indeed, the 2009 NSS enumerates several such threats. It states that "the national interests of the RF in the long term consist of the following: developing democracy and civil society, and the enhancement of the competitiveness of the national economy; ensuring the solidity of the constitutional system, territorial integrity, and sovereignty of the RF; transforming the RF into a world power, whose activity is directed at supporting the strategic stability and mutually beneficial partner relationships within the multi-polar world."[33] In addition to externally generated threats in the sections on "national defense" and "strategic stability and equitable strategic partnership," the document identifies internal threats and associated security objectives in sections on "state and public security," "improvement of the quality of life of Russian citizens," "economic growth," "science, technology and education," "healthcare," "culture," and "the ecology of living systems and environmental management."

The greatest internal threat to Russia's security has originated in Chechnya and the neighboring republics of the northern Caucasus. Violent incidents in this region persist despite Moscow's stepped-up security measures. More dramatically, numerous terrorist attacks, attributable to Islamic groups seeking separation, have periodically reached Moscow and other distant areas. The issue of military use to quell such internal political disturbances—increasingly sensitive during the Gorbachev years in minority republics such as Georgia and the Baltic States—became acutely controversial in the period surrounding the October 1993 confrontation between President Yeltsin and the parliament. Although Russian military authorities reportedly resisted, the acceptability of this practice was incorporated in the new military doctrine that was adopted by the Security Council and the president in November 1993. The document spelled out the circumstances under which the military could be activated in internal conflicts: "Proceeding from the premise that the security of Russian Federation citizens and the state as a whole is paramount, the doctrine provides for the use of Army units, Ministry of Internal Affairs internal troops, border troops and other forces in situations in which all other means and capabilities have been exhausted." (The National Security Doctrine of 2000 further broadened the circumstances under which military force could be used inside the country, to allow it "in the event of emergence of a threat to citizens' lives and also of violent changes to the constitutional system."[34])

In addition to the "traditional missions" of the armed forces in "deterring and repulsing aggression," the doctrine envisioned an additional mission: "conducting peacekeeping operations within the CIS and, by decision of the UNSC and other international bodies, outside the Commonwealth provided they are not in conflict with Russian's interests and Russian law." The possibility of maintaining Russian troops and bases on the territory of other countries was not ruled out, "on the basis, of course, of bilateral and multilateral agreements with the countries in which they would be located." Seeing the chief danger to peace in local wars, the doctrine emphasized the need to create mobile forces that could be dispatched quickly to danger spots.[35]

As for the danger of nuclear war, it was said to be considerably reduced, and nuclear weapons were viewed, in then Defense Minister Grachev's words, "primarily as a political means of deterring aggression, not as a means to conduct military operations." Accordingly, the military doctrine reversed Gorbachev's earlier pledge that the country would not be the first to use nuclear weapons. That declaration, now identified as mainly a "propaganda thesis," was said to have justified enormous conventional forces and a quest for attaining a superior nuclear potential. These military attributes were no longer necessary in the new circumstances, in which Russia "does not regard any country as its adversary." However, in order to discourage other CIS members from retaining or acquiring nuclear weapons, the Russian doctrine extended the "nuclear umbrella" to all members of the CSTO "and guarantees their territories against any invasion by any country that is an ally of any nuclear power."

Putin enunciated a "new military doctrine" when he spoke to the top military brass in October 2003. He expressed confidence that Russia had sufficient strategic missile strength—pressing into service previously stockpiled SS-19 missiles with multiple warheads to replace older missiles—to overcome any potential American antimissile system.[36] And though the Bush-style "national missile defense" project was shelved by the Obama administration in 2009, the plans of the United States and NATO to erect a more limited missile defense system have continued to elicit alarmist statements from Russian generals and politicians.

In the same 2003 statement, Putin stressed that Russia's military command would be paying far more attention in its troop training to peace operations, special operations, antiterrorist actions, and local wars. On the whole, in its emphasis on mobile units and peacekeeping responsibilities, especially in the CIS, the Russian military doctrine clearly envisaged the armed forces as an active instrument of the country's foreign policy.

Efforts over the past decade and a half to achieve a "military reform," featuring an all-volunteer force, less top-heavy with officers and more mobile, have advanced very little. Accordingly, the emphasis in Russia's military doctrine is on maintaining an adequate strategic deterrent, as well as on creating political conditions on its border to minimize threats. According to the NSS, "The strategic goals related to improving national defense consist of preventing global and regional wars and conflicts and likewise of realizing strategic deterrence in the interests of ensuring the country's military security."[37]

Recovery of Russia's military capabilities toward the end of Putin's first term as president was certainly being aided by the recovery of its economy, fueled by soaring revenues from petroleum exports. President Dmitri Medvedev announced in 2010 that the defense budget would be set at 2.8 percent of GDP for the next decade. At that level, it would remain below the proportion (3.5 percent) once targeted by Yeltsin.[38] And although the declared defense budget is somewhat misleading, given the existence of large numbers of armed personnel in other government agencies—especially troops of the Ministry of Interior and of the FSB—total military expenditures, estimated to be about $72 billion in 2011, still remain at less than 10 percent of the U.S. defense budget and far below the level of expenditure in the Soviet years.

However, the increase has allowed a certain amount of catch-up, especially in expenditures on research and development and acquisition of military weaponry—but there was much more catching up required if the appetites of military commanders were to be satisfied. In 2004, military expert Alexei Arbatov noted, "less than twenty percent of the weapons that troops now have are relatively new—purchased less than ten years ago." And he complained that real progress in reequipping the armed forces and sustaining the Russian defense industry would not come until the size of the military had been reduced to a level well below the announced target of 1 million. In Arbatov's words, "no reasonable budget can sustain an army consisting of 1.2 million men. And there is no need for such an army. After all, we are not going to start another Great Patriotic War."[39] By 2012, the size of the armed forces had declined to just over 1 million men—still much larger than Arbatov and other experts felt was needed.

Together with the military, the KGB was another "power-wielding" institutional actor that regularly found a place at the table when foreign policy decisions were made in the Soviet Union. It continues to do so, without having undergone significant reform, in democratizing Russia. The FSB is the chief successor to the internal functions of the KGB, while the SVR, the Foreign Intelligence Service, has taken over the external intelligence functions. Vladimir Putin, who left the KGB as a Lieutenant-Colonel, briefly headed the FSB before assuming higher office. Yevgeny Primakov, who served Yeltsin as both foreign minister and prime minister, had previously led the SVR. Although Putin has gained attention for elevating security service alumni (the *siloviki*) into high positions, it was actually the ostensibly more "democratic" Yeltsin who began the trend; his last three Security Council secretaries and last three prime ministers had spent most of their careers in the security or intelligence services.

Reflections

Much of the commentary in the Russian press in 2011 centered around the forthcoming 2012 presidential election and the question of whether Dmitri Medvedev would seek another term or whether his patron, Prime Minister Vladimir Putin, would return to the Kremlin as president. Some observers, discerning a more "pro-Western" and "liberal" orientation in Medvedev's statements and actions, argued that Putin's return would bring about an end to the more cooperative relationship with the United States ("reset") that had prevailed since 2009. Many others, however, considered such views as misperceptions —confusing mere differences of style with actual differences in policies. When it was announced, in September 2011, that Putin would seek to return to the presidency for a third term, it was clear that Putin had made the decision, and that any candidates put forward by the opposition political parties—whether the communists on the left or the liberals on the right—were almost certainly going to be defeated. What was not foreseen, however, was that the blatant falsification of results of the December 2011 parliamentary elections by functionaries of the ruling United Russia Party would evoke such strong popular demonstrations in the streets of Moscow and St. Petersburg. Putin's victory the following March did not appear to require such a degree of manipulation, but his return to the Kremlin was clearly less than triumphal, given the growing discontent over the political system he had put in place. Nevertheless, Vladimir Putin remains Russia's most popular politician—a product not merely of his success in restoring an aura of relative stability and economic growth, but also of the credit that he has been given for restoring self-confidence and stature to Russia's role in the world.

Notes

1. Robert H. Donaldson and Joseph L. Nogee, *The Foreign Policy of Russia: Changing Systems, Enduring Interests*, 4th ed. (Armonk, N.Y.: M.E. Sharpe, Inc., 2009), pp. 107–109.

2. For an insider's analysis, see Yegor Gaidar, *Collapse of an Empire: Lessons for Modern Russia*, trans. Antonina W. Bouis (Washington, D.C.: Brookings Institution Press, 2007).

3. For details, see Karen Dawisha and Bruce Parrott, *Russia and the New States of Eurasia: The Politics of Upheaval* (Cambridge, UK: Cambridge University Press, 1994).

4. As presidential advisor Sergei Stankevich wrote in March 1992: "Foreign policy with us does not proceed from the directions and priorities of a developed statehood. On the contrary, the practice of our foreign policy will help Russia become Russia." Sergei Stankevich, "A Power in Search of Itself," *Nezavisimaia gazeta*, March 28, 1992, quoted in James Richter, "Russian Foreign Policy and the Politics of National Identity," in *The Sources of Russian Foreign Policy after the Cold War*, Celeste A.Wallander, ed. (Boulder, Colo.: Westview, 1996), p. 69.

5. Mark Kramer, "What Is Driving Russia's New Strategic Concept?," Program on New Approaches to Russian Security (PONARS), Harvard University, January 2000, in *Center for Defense Information (CDI) Russia Weekly*, no. 85. The final version of the concept was published in the military weekly, *Nezavisimoe voennoe obozrenie*, January 14, 2000, in *Johnson's Russia List*, no. 4072; "Foreign Policy Concept of the Russian Federation," *Rossiiskaia gazeta*, July 11, 2000, in *Johnson's Russia List*, no. 4403, July 14, 2000; Andrei Shaburkin, "Russia Changes Its Military Doctrine," *Nezavisimaia gazeta*, February 5, 2000, in *Current Digest of the Post Soviet Press (CDPSP)* 52, no. 6 (2000), pp. 13–14.

6. Viktor Litovkin, "Security Is Best Achieved through Coalition," *RIA Novosti*; Vladimir Isachenkov, "Putin Says Russian Military Still Mighty"; "Putin Says RF Having Big Amounts of Heavy Missiles," *ITAR-TASS*, in *CDI Weekly*, no. 276, October 2, 2003; "President Putin's Address at the Plenary Session of Ambassadors and Permanent Representatives of Russia," *RIA Novosti*, in *Johnson's Russia List*, no. 8290, July 13, 2004.

7. Text of the revised "Foreign Policy Concept of the Russian Federation," at http://www.maximsnews.com/news20080731russiaforeignpolicyconcept10807311601.htm.

8. Natsional'naia bezopastnost' Rossii, "Strategiia natsional'noi bezopastnosti Rossiiskoi Federatsii do 2020 goda," at http://www.scrf.gov.ru/documents/11/99.html, translated at http://rustrans.wikidot.com/russia-s-national-security-strategy-to-2020. Hereafter, references to translated version as NSS, with indication of section and paragraph.

9. See Harley Balzer, "Vladimir Putin's Academic Writings and Russian Natural Resource Policy," and V. V. Putin, "Mineral Natural Resources in the Strategy for Development of the Russian Economy," in *Problems of Post-Communism*, vol. 53, no. 1 (January-February 2006), pp. 48–54. There have been a number of reports that claim that much of Putin's "thesis" was plagiarized from two University of Pittsburgh economists, but since the document itself is not publicly available, definitive proof is lacking.

10. NSS, II-11.

11. NSS, II-9.

12. Dale Herspring, "Russia's Crumbling Military," *Current History*, October 1998, p. 325.

13. As the Russians see it, as well as certain outside observers, there is no difference between a forceful separation of Kosovo from Serbia in the wake of "ethnic cleansing" and the use of force to gain independence from Georgia for Abkhazia and South Ossetia, which were also victims of ethnic repression. See, for example, Oksana Antonenko, "From Kosovo to South Ossetia: in search of a precedent," RIA Novosti, August 19, 2008, at http://en.rian.ru/analysis/20080819/116126973.html.

14. Angela Charlton, "Putin Says Russia Could Join NATO," *The Independent*, March 5, 2000, at http://www.independent.co.uk/news/world.eur/putin-says-russia-could-join-nato-722225.html.

15. Robert H. Donaldson, "Alternative Scenarios for a Pan-European Security System," paper prepared for the 52nd Annual Convention of the International Studies Association, Montreal, Quebec, Canada, March 16–19, 2011.

16. Anne L. Clunan, *The Social Construction of Russia's Resurgence: Aspirations, Identity and Security Interests*, Baltimore: Johns Hopkins University Press, 2009.

17. Martin A. Smith, *Russia and NATO since 1991: From Cold War through Cold Peace to Partnership?* (London and New York: Routledge, 2006), p. 109.

18. Sergei Karaganov and Timofei Bordachev, "Towards a New Euro-Atlantic Security Architecture," Report of the Russian Experts for the Valdai Discussion Club Conference, London, December 8–10, 2009, p. 6.

19. Donaldson and Nogee, p. 263.

20. Donaldson and Nogee, pp. 112–113 and footnote 9.

21. Donaldson and Nogee, pp. 179–181.

22. Paul Reynolds, "New Russian World Order: The Five Principles," BBC News, September 1, 2008, at http://news.bbc.co.uk/2hi/europe/7591610.stm.

23. NSS, II-13–15.

24. Donaldson and Nogee, pp. 323–324.

25. Ibid., pp. 285–290.

26. "Russian President Visits Disputed Kuril Islands," BBC News Asia-Pacific, November 1, 2010, at http://www.bbc.co.uk/news/world-asia-pacific-11663241.

27. Donaldson and Nogee, p. 282.

28. See, for example, the September 2010 issue of *Current History* on "China and East Asia"—you would not know from it that Russia was in East Asia.

29. Donaldson and Nogee, pp. 279–280.

30. "Prime Minister Vladimir Putin Meets with Participants of the 7th Meeting of the Valdai International Discussion Club in Sochi," at http://premier.gov.ru, translated at *Johnson's Russia List*, no. 171, September 8, 2010.

31. For detailed discussion of this issue, see Donaldson and Nogee, pp. 296–317.

32. A more complete description of the history of the Security Council may be found in Donaldson and Nogee, pp. 126–132.

33. NSS, III-21.

34. See Donaldson and Nogee, pp. 142–144.

35. Vasili Kononenko, "Russian Military Doctrine Lacks No-First-Use Pledge," *Izvestia*, November 4, 1993; Pavel Felgengauer, "Russia Shifts to Doctrine of Nuclear Deterrence," *Segodnia*, November 4, 1993; Roman Zadunaiski, "Military Doctrine Adopted," *Rossiiskie vesti*, November 5, 1993, in *CDPSP* 45, no. 44 (1993), pp. 11–12.

36. Litovkin; Isachenkov; in *CDI Weekly*, no. 276, October 2, 2003.

37. NSS, IV-1.

38. Donaldson and Nogee, p. 145.

39. "Inflated Ranks, Tiny Budget and Enduring Paranoia," *Mosnews.ru/Gazeta.ru*, August 24, 2004, in *Johnson's Russia List*, no. 8341, August 24, 2004.

PART V

POTENTIALLY (RE-)EMERGING WORLD

12 To Survive or Lead? The Two Sides of Nigeria's National Security Strategy

Jon Hill

NIGERIA'S NATIONAL SECURITY STRATEGY CONTINUES TO BE shaped by two dominant impulses. The first is self-preservation. Its federal government is currently fighting two counter-insurgency campaigns: one against *Boko Haram* in the north and the other against the Movement for the Emancipation of the Niger Delta (MEND) in the south. Through their actions, both groups are destabilizing the country and threatening its long-term survival. The second impulse is to turn Nigeria into a regional power, to establish it as sub-Saharan Africa's preeminent state, to make it the international community's preferred and essential African partner.

While these impulses are not wholly incompatible, they sit uneasily alongside one another. Clearly securing the government and the people against these domestic threats is Abuja's first priority. Yet its pursuit, or, more precisely, the fact that it is necessary, embarrasses the federal government and raises serious questions about the viability of its leadership ambitions. Just how much attention can Abuja pay to continental issues when confronted with such dangers at home? What resources can it dedicate to resolving problems elsewhere in Africa? Will other governments listen to what it has to say if it is unable to safeguard its own authority? And what moral leadership can it provide if its own conduct in combating these insurgencies is constantly called into question?

This tension continues to present Washington, London, Paris, and Brussels with an equally difficult dilemma. On the one hand, they are keen for

Nigeria to pursue its leadership ambitions if it means that the country will continue to play an active part in helping to resolve Africa's problems. Yet on the other, none of these functions is nearly as important as securing Nigeria against *Boko Haram* and the MEND, and ensuring that the country remains intact. Indeed, Nigeria's collapse and disintegration would not only create a crisis that dwarfs nearly all those currently plaguing the continent, it would greatly exacerbate many of these other problems. In short, the federal government must put Nigeria's own security and survival first.

The aim of this chapter is to examine this tension and its implications for Nigeria and Africa. For if the threat from *Boko Haram* continues to grow as rapidly as it has over the past three years, the federal government may be forced to choose between its obligations at home and its ambitions abroad. Either way, the consequences for the international community may be profound. To do this, the chapter is divided into three main parts. In the first it provides a brief overview of Nigeria's national security apparatus. In the second it sets out Nigeria's importance to the international community and leadership ambitions. And in the third it looks at the main domestic threats currently confronting the federal government.

Nigeria's National Security Strategy

Nigeria does not have a permanent National Security Council (NSC). Instead, the duties are performed collectively by the president's National Security Adviser (NSA), the service chiefs, and the Director General (DG) of the State Security Service (SSS). Formally, it falls to the NSA to draw together and manage the advice and guidance offered by the heads of these various agencies. Yet in practice, given the personal nature of political power in Nigeria and the limits placed on the NSA's authority, the president consults with the DG and chiefs of the army, navy and air staffs directly since the NSA has no meaningful power over their military and SSS counterparts.

Ostensibly, in an effort to improve this working relationship, the president usually appoints an NSA who has close ties with the military. Indeed, the current NSA, Sambo Dasuki, is a retired colonel.[1] His three immediate predecessors—Andrew Azazi, Aliyu Mohammed Gusau, and Abdullahi Sarki Mukhtar—were once Chief of Defence Staff, Chief of Army Staff, and Commander of the First Division of the Nigerian Army, respectively.[2] It is perfectly possible that this arrangement helps keep the wheels of national security planning and decision-making oiled. Yet it is also the result of other political and historical considerations springing from the long years the military ruled Nigeria. Since the restoration of civilian rule in May 1999, the position of NSA has been used, in part, to help reconcile the armed forces to their new, nongoverning role.

In truth, the various officials and agencies involved in devising Nigeria's national security strategy are broadly the same as those who sit on the U.S. and British NSCs. Yet arguably, their collective deliberations are conducted in a more ad hoc fashion. The absence of a more formal structure, in the form of a council, need not make the Nigerian process of discussing and developing a national security strategy any less effective. Yet it does seem to prevent the compilation and publication of a single document outlining the country's main strategic challenges and objectives. Unlike either the United States or the UK, Nigeria does not produce a collated NSS for public consumption. It is highly likely that no such document exists for private consultation either.[3]

A Leader for and in Africa

Nonetheless, it is still possible to piece together the core tenets of Nigeria's national security strategy from what successive administrations have said and done. Their desire to establish Nigeria as a leader both in and for Africa has long been clear. Today, Nigeria's importance and influence rests on four main pillars—people, petroleum, politics, and peacekeeping. Nigeria is important because of the size of its population. Regardless of whether its populace is 150 million people (as estimated by the U.S. Department of State in 2010),[4] 162 million people (as estimated by the UN in 2011),[5] or 167 million people (as estimated by the Nigerian government in 2012),[6] Nigeria is still the most populous country in Africa.[7] And its population will continue to grow rapidly over the coming decades, rising to an estimated 400 million people by 2050 and 750 million people by 2100.[8] Nigeria is important simply because so many people live there.

Nigeria is one of the world's most important energy providers, even though the data on its oil and gas industries is as contested as that of its population. The Nigerian National Petroleum Corporation (NNPC) states that the country has proven reserves of 28.2 billion barrels. But the Organization of the Petroleum Exporting Countries (OPEC) and the U.S. Department of Energy (DOE) both believe it has 37.2 billion barrels. The NNPC claims that

Nigeria extracts up to 2.5 million barrels of oil each day, but the OPEC and the U.S. DOE say that it pumps 2 million, and between 1.7 and 2.1 million barrels a day, respectively. OPEC maintains that the country exports around 2.4 million barrels of crude oil each day, but the DOE places the figure lower, at 1.8 million barrels a day.[9]

Such divergences can also be found in the latest data on Nigeria's gas sector. The NNPC avows that the country has proven reserves of 165 trillion cubic feet. But OPEC and the DOE state that it has reserves of 180 trillion and 187 trillion cubic feet, respectively. OPEC insists that the country pumps 992 billion cubic feet of gas per day, but the DOE says that it only extracts 820 billion cubic feet per day. And OPEC maintains that the country exports 706 billion cubic feet of gas each day, but the DOE states that it ships no more than 565 billion cubic feet daily.[10]

Even though there is little consensus on how much oil and gas Nigeria has, extracts, and exports, there is hardly any disagreement over the importance of these industries to the country's economy or the global energy markets. To begin with, it is difficult to overstate the extent of Nigeria's reliance on oil money, the revenue it earns both from exports, and the rents it charges the multinational companies that drill within its borders. Indeed, both the DOE and the World Bank estimate that oil brings in 95 percent of the country's foreign currency earnings[11] and 40 percent of the Nigerian federal government's total income.[12]

The 1.7 to 2.5 million barrels of oil Nigeria pumps every day represents around 2 percent of the total amount extracted worldwide. Though it contributes far less to global supply than Russia or Saudi Arabia, Nigeria is still the tenth largest oil producing country in the world. Most of what it extracts, or between 1.8 and 2.4 million barrels, is sold abroad.[13] Half of these exports are bought by the United States and represent around 8 percent of the total amount the country imports daily.[14] Nigeria, then, not only plays an important role in maintaining the global supply of oil, so vital to economic growth and political stability, but it is also a major supplier to the United States directly.

Nigeria's importance as an energy provider is raised still further by the quality of the oil it produces. First, its oil is described by industry experts as light and sweet. This means that it is ideally suited to refinement into motor fuels. Second, the country's close proximity to the Atlantic sea lanes facilitates the shipment of its oil to refineries in North America and Europe. And third, the on-going disruption of supplies from other important oil-producing countries in North Africa and the Middle East means that it is more important than ever to maintain the flow from Nigeria.[15] The country's oil is one of the main reasons why a majority of Nigerian and international politicians want the country to remain together.[16]

Nigeria is simply too important for the international community to ignore. Its dominance of the West Africa subregion is beyond both question and challenge. The size of its population gives it unparalleled political, military, economic, and cultural influence. And its preeminence there automatically makes it one of Africa's core powers. As such, the country has an important role to play in shaping the African Union—how it is organized, the way it conducts its business, what it debates, and the decisions it makes—and the continent's responses to global events. In short, Nigeria's position within Africa means that the international community must consider its opinions and seek, whenever possible, to work with it.

Nigeria has made vital contributions to a range of peacekeeping missions throughout Africa. At the behest of the United Nations, the African Union, the Economic Community of West African States (ECOWAS), and individual foreign governments, it has supported:

- ONUC (Republic of the Congo, 1960—1964)
- UNOSOM I (Somalia, 1992)
- UNITAF (Somalia, 1992—1993)
- UNOSOM II (Somalia, 1993—1995)
- ECOMOG/UNOMIL (Liberia, 1990—1997)
- ECOMOG/UNOMSIL/UNAMSIL (Sierra Leone, 1991—2005)
- MONUSCO (Democratic Republic of the Congo, 1999—present)
- UNMIS/AMIS/UNAMID (Sudan/Dafur, 2005—present)
- AMISOM (Somalia, 2007—present)

These deployments, both individually and collectively, represent a huge military investment by the country; tens of thousands of troops, thousands of dead and wounded, and hundreds of millions of dollars.

Yet there is more to Nigeria's contributions than just the soldiers it has sent and the arms it has committed. For time and again and at crucial moments it has shown both leadership and resolve. Certainly it was one of the key driving forces behind the missions to Liberia and Sierra Leone. As the founder and mainstay of ECOWAS, it played a pivotal role in persuading the organization's other members to deploy monitoring groups (ECOMOG) to each country.

Moreover, and just as crucially, it has been willing to commit its forces when others (most notably South Africa, the United States, Britain, France, and the European Union) have not. In doing this, Nigeria has made its strongest claim to be sub-Saharan Africa's preeminent state.

At present, this title undoubtedly belongs to South Africa, and certainly Abuja sees Pretoria as its principal rival. Relations between the two countries have improved significantly since the end of white minority rule in 1994, but there are still some notable points of friction. The anti-foreigner riots in South Africa in the summer of 2008 were mainly aimed at the hundreds of thousands of Zimbabwean refugees living there[17] but also targeted the sizeable Nigerian community, which is widely seen as a major source of crime. These suspicions were clearly expressed in the internationally acclaimed 2009 film *District 9*,[18] and only appeared to be confirmed by Henry Okah's arrest in Johannesburg the following year.[19]

Since the end of apartheid, South Africa's own NSS has been heavily revised and its security forces completely reorganized. Instead of undertaking police actions in townships or conducting counter-insurgency operations in the bush along its northern border, South Africa's armed forces have been remodeled to carry out UN and AU approved peacekeeping operations in Africa. Yet so far, mainly because of this post-apartheid introspection and reorganization, the country has contributed only around 3,400 peacekeeping troops to just a handful of missions. South Africa's largest deployments to date have been to Burundi and the Democratic Republic of the Congo (DRC).[20] It has not committed the same number of troops or engaged in the same number of missions as Nigeria has.

In contributing to so many peacekeeping missions on the scale that it does, Nigeria emphasizes a way in which it is of greater use to Africa and the international community than South Africa. Indeed, North American, European, and other governments are extremely anxious that Nigeria continues to perform this role, not least because of their unwillingness to send their own forces. Indeed, since its disastrous involvement in UNOSOM II mission, the United States has been extremely reluctant to deploy its forces either on peacekeeping missions or to Africa. Washington now looks to others to meet the UN's frequent requests for troops. Nigeria, by virtue of its willingness to supply at least some of them, enables the United States to avoid committing its own forces, which are currently deployed elsewhere, on operations that command little domestic support yet also meet its broader strategic objectives for Africa.

Nigeria plays a vital role in providing the military presence that the United States, Britain, and others want but are unwilling to supply themselves. In so doing, Abuja not only makes clear its leadership ambitions but also furthers them. This, of course, requires it to say yes to such initiatives far more frequently than it says no. And by getting involved, Abuja should be able to achieve a presence in parts of the continent that historically have been beyond its sphere of influence. Sudan and Somalia, for example, are both places with far stronger regional ties to Kenya than Nigeria.

The Home Front

Mercifully for its hard-pressed government and public, Nigeria has never been seriously threatened by outside invasion and conquest. It has, of course, fallen out with some of its neighbors on various occasions, most notably Cameroon over the sovereignty of the Bakassi Peninsula,[21] Ghana over the fate of its citizens living in Nigeria,[22] and São Tomé e Príncipe over the position of their shared maritime border.[23] Yet despite the bitterness and bad faith that sometimes marked these rows, they rarely resulted in anything more serious than the odd, small-scale skirmish along the border in question. Indeed, this part of the continent has been spared much of the interstate violence and bloodshed that has so shaped the Horn of Africa's development since independence.

The major threats to Nigeria's security both today and in the past have come from within. At present the federal government is grappling with two distinct insurgencies. Over the past three years *Boko Haram* has been the main focus of international attention. Its growth in prominence can be directly traced through a series of key events. The group first drew significant international attention in late July 2009 when its militants fought street battles with the police and army in the north-eastern city of Maiduguri.[24] This attention was renewed on Christmas Eve 2010 when the group carried out simultaneous bomb attacks on several targets, including two churches, in the city of Jos. The synchronized nature of this assault, timing, and high casualty rate earned it headlines all over the word.

On Christmas Day 2011 it again bombed churches in the towns of Madala, Jos, and Gadaka, and suicide bombed the SSS station in the town Damaturu.[25] But arguably the group's most spectacular action to date, the one that has gained it the most attention, came in August 2011 when it suicide-bombed the UN in-country headquarters in Abuja.[26] This attack sent an especially chill wind coursing through the capitals of Europe and North America. Not only was this the first suicide bombing ever carried out in Nigeria, but it also represented a major assault on the international community and raised terrifying questions about the group's capabilities. For if it could bomb the UN in one of the most secure compounds in the country, then would it be able to attack the U.S. embassy, British High Commission, and other foreign missions too?

Indeed, it was in response to this attack that the U.S. House of Representatives Subcommittee on Counterterrorism and Intelligence drafted and published its report *Boko Haram: Emerging Threat to the U.S. Homeland.* Yet is *Boko Haram* really that dangerous? Is the amount of international attention paid to it truly warranted? Certainly *Boko Haram* is not the only armed group currently operating in Nigeria. Far from it, in fact, as the Nigerian government continues to do battle with a range of armed factions in the far south of the country, including the MEND. And in some ways, these groups pose a far more serious and immediate threat to the interests of the Nigerian government and those of the international community than does *Boko Haram.*

To begin with, the MEND continues to target parts of Nigeria's oil infrastructure. The damage it causes to pipelines and wells, and the oil it steals or bunkers from them, is depriving Abuja of tens of millions of dollars in much needed income and forcing up prices on the international market.[27] Then there is the threat to foreign nationals working in the region and international shipping off the cost of Nigeria as the MEND continues to kidnap those working in the oil industry and to attack oil and other vessels sailing in the Gulf of Guinea. Indeed, Nigeria's coastal waters are now among the most hazardous in the world.[28] Finally, the MEND has carried out its own mass casualty strikes such as its bombing of the Ministry of Justice headquarters in Abuja in October 2010. So does *Boko Haram* really pose a greater danger to Nigeria and the international community than the MEND? Does it represent a qualitatively different and more serious type of threat?

Part of the concern over *Boko Haram* is undoubtedly linked to the existence and on-going activities of the MEND. The threat presented by *Boko Haram* compounds that posed by this and the other Niger Delta factions. But there are other reasons why the international community is currently more fearful of *Boko Haram* than it is the MEND. Foreign governments' are particularly concerned by the speed and scale of *Boko Haram*'s evolution over the past three years.[29] During this time, the group has undergone six important transformations. To begin with, there are its goals. Gone is its desire to live in peaceable isolation. Since the Battle of Maiduguri, *Boko Haram* has proclaimed and pursued three more dynamic objectives. The first is to ensure that sharia law is more fully enforced in the North.[30] The second is to extend sharia law to the whole of the country including those areas inhabited mainly by Christians. And the third is to crack down on corruption.

The second transformation undergone by *Boko Haram* has been its use of violence. Originally the group mostly resembled a religious commune. It launched some armed assaults, most notably in December 2003 and in September and October 2004, but these were infrequent, together resulted in the deaths of tens of people, and were made only against the security forces. They were nowhere near the scale and intensity of the gun battles it fought against the police and army in July 2009 in Maiduguri and its environs. Since December 2010 hardly a month has passed without an attack being carried out. While *Boko Haram* might not be responsible for everything that is laid at its feet, it is certainly responsible for a great deal. Indeed, since the start of 2010, the group is thought to have killed more than a thousand people.[31]

This sharp rise in the death toll is partly a result of *Boko Haram*'s expansion of its list of targets. The third transformation the group has undergone, therefore, is its identification of a growing range of enemies. Most of *Boko Haram*'s victims are still police officers and military personnel. Yet since the Battle of Maiduguri, it has also attacked politicians and civil servants, religious leaders and community elders, Muslims engaging in what it considers to be un-Islamic behavior, Christians, students, foreign nationals, and international organizations. Indeed, there are now very few groups or sections of society that it is not prepared to strike.[32]

Boko Haram's targeting of at least some of these new groups has both been encouraged and made possible by its expansion of its area of operations. Indeed, the fourth transformation it has undergone has been its movement into parts of the country in which it was previously unknown. Certainly the north-eastern states, and in particular the city of Maiduguri, remain its prime area of activity. But over the past three years it has successfully carried out attacks as far south as Abuja, which is located at Nigeria's geographic centre. Now the whole of the north is open to its fighters, who continue to mount dozens of assaults right across the region. And so too, it would seem, are Nigeria's neighbors, with hundreds of *Boko Haram* militants recently spotted in nearby Mali.[33]

Another important reason for the rising death toll is *Boko Haram*'s ability to mount a growing range of attacks. Indeed, the fifth transformation it has undergone has been the development of its operational capabilities, or what it can actually do. Most of the assaults carried out by the group still fall into one of three categories—targeted assassinations of prominent public figures; non-suicide bombings of specific locations; and gun attacks on groups of people and particular places. Yet it is also able to conduct many other types of operations, including prison breakouts, suicide bombings, ambushes, and full-frontal assaults. The group's repertoire has expanded greatly since the Battle of Maiduguri.[34]

And finally the sixth transformation undergone by *Boko Haram* is its growing links to other terrorist groups, most notably Al-Shabaab based in Somalia and, in particular, Al Qaeda in the Land of the Islamic Maghreb (AQLIM), based in Algeria.[35] To be clear, *Boko Haram* has always had international ties. Many of the young men who joined it initially had recently returned from Sudan where they had been completing their religious studies.[36] Its membership has always included a good number of Nigerians, Chadians, and Cameroonians. Moreover, by embracing Salafist beliefs and ideas, it embraced a global doctrine that has, at its heart, a will to universalism, a desire to spread and propagate itself.

But in the early days its links to groups like AQLIM were fairly weak. Indeed, it was not until the Battle of Maiduguri and, more importantly, the execution of *Boko Haram*'s leader, Mohammed Yusuf, during the battle that these connections were forged in earnest. As a result of Yusuf's death and the crackdown the security forces launched against the group, many of its surviving senior members fled to neighboring Niger and Chad, where they received money, weapons, training, and other help from AQLIM agents sent to assist them by their commander-in-chief, Abdelmalek Droukdel. In recognition of this support, *Boko Haram*'s then head, Abu Bakr Bin Muhammed al Shakwa, swore an oath of allegiance to Droukdel in October 2010.

Boko Haram's link-up with AQLIM helps explain at least some of these transformations, most notably the expansion of its targets and operational capabilities. Indeed, any doubts as to the extent and significance of AQLIM's influence on *Boko Haram* were emphatically dispelled by its successful assault on the UN headquarters. Not only was this the first suicide bombing in Nigeria's history, but to carry it out the group had to develop a range of capabilities it had not previously possessed and which, just as crucially, it had evinced little desire to acquire prior to the battle of Maiduguri. Finally, the three hundred or so *Boko Haram* fighters recently seen in Mali are almost certainly there because AQLIM asked them to be.

Reflections

In all but the rarest of cases, national security strategies are multifaceted. Habitually, they look both inwards and outwards as political leaders and security providers grapple with the dangers emanating from both within and without. Nigeria's NSS is no exception. At its heart lie two core ambitions—to achieve that which the country must, while pursing that which it would like. Unsurprisingly, a great deal of Nigeria's NSS is devoted to identifying and dealing with the threats that pose the gravest dangers to the country's stability, territorial integrity, and the safety and economic future of its citizens. The most serious of these challenges are presented by the MEND and *Boko Haram*. And to meet them, the federal government is compelled to invest significant military and police resources. In fact, Abuja has long made its greatest military intervention within its own borders.

Yet the federal government is also pursuing its goal of establishing Nigeria as sub-Sahara Africa's leading voice. This is an old ambition and remains intimately bound to Nigerian views of what type of role they would like their country to play on the international stage. Supporting this claim are a range of important pillars, including the size of Nigeria's population, its control of significant oil and gas reserves, its access to the great revenue their sale abroad generates, its hegemonic position in West Africa, and its active support of various peacekeeping missions around the continent. Its achievement though, is partially dependent on the federal government's success in maintaining order at home. For this stability directly impacts on Abuja's ability to commit its military forces elsewhere. And certainly Washington, London, Paris, and Brussels are hoping that it can meet these challenges.

Notes

1. Dasuki was appointed to the post on 22 June 2012.

2. This was Gusau's second term as NSA, having previous served President Obasanjo from May 1999 to June 2006. Gusau's most recent stint as NSA lasted from March 2010 to April 2010, and Mukhtar held the position from May 2006 to March 2010.

3. This is the considered opinion of a senior member of the British High Commission Abuja, interview by author, Oxford, 9 November 2010.

4. U.S. Department of State, *2010 Human Rights Report: Nigeria*, 8 April 2011, available at http://www.state.gov/j/drl/rls/hrrpt/2010/af/154363.htm (accessed 6 July 2012).

5. United Nations Development Programme, *Nigeria Country Profile: Human Development Indicators*, 2011, available at http://hdrstats.undp.org/en/countries/profiles/NGA.html (accessed 6 July 2012).

6. Permanent Mission of Nigeria to the United Nations, *Statement by Jamin Dora Zubema Director-General National Population Commission on Adolescents and Youths (Nigeria Experience) at the 45th Session of the United Nations Commission on Population and Development (UNCP&D)*, 24 April 2012, available at http://www.un.org/esa/population/cpd/cpd2012/Agenda%20item%204/Country%20statements/Nigeria_Item4.pdf (accessed 6 July 2012).

7. The next most populous African country is Ethiopia. Its population currently stands at 84 million. U.S. Department of State, *Background Note: Ethiopia*, 2 April 2012, available at http://www.state.gov/r/pa/ei/bgn/2859.htm (accessed 6 July 2012).

8. These estimates are the medium variants identified by the UN. United Nations Department of Economic and Social Affairs, *Country Profile: Nigeria*, available at http://esa.un.org/unpd/wpp/country-profiles/country-profiles_1.htm (accessed 6 July 2012).

9. All these figures are drawn from the following sources: Nigerian National Petroleum Corporation, *Oil Production*, 23 February 2012, available at http://www.nnpcgroup.com/NNPCBusiness/UpstreamVentures/OilProduction.aspx (accessed 23 February 2012), Organisation of Petroleum Exporting Countries, *Nigeria*, 2012, available at http://www.opec.org/opec_web/en/about_us/167.htm (accessed 23 February 2012) and U.S. Department of Energy, *Country Analysis Briefs: Nigeria*, August 2011, available at http://www.eia.gov/EMEU/cabs/Nigeria/pdf.pdf (accessed 23 February 2012).

10. The figures provided by OPEC were in cubic meters. They were converted into cubic feet to facilitate their comparison with those provided by the NNCP and the U.S. DOE. The amount of gas the U.S. DOE believes Nigeria exports is determined by subtracting the amount Nigeria consumes each year from that which it produces. Nigerian National Petroleum Corporation, *Oil Production*, 23 February 2012, available at http://www.nnpcgroup.com/NNPCBusiness/UpstreamVentures/OilProduction.aspx (accessed 23 February 2012), Organisation of Petroleum Exporting Countries, *Nigeria*, 2012, available at http://www.opec.org/opec_web/en/about_us/167.htm (accessed 23 February 2012) and U.S. Department of Energy, *Country Analysis Briefs: Nigeria*, August 2011, available at http://www.eia.gov/EMEU/cabs/Nigeria/pdf.pdf (accessed 23 February 2012).

11. World Bank, *Nigeria: Country Assistant Strategy*, 2011, available at http://web.worldbank.org/WBSITE/EXTERNAL/COUNTRIES/AFRICAEXT/NIGERIAEXTN/0,,menuPK:368909pagePK:141132piPK:141105theSitePK:368896,00.html (accessed 23 February 2012) and U.S. Department of Energy, *Country Analysis Briefs: Nigeria*, August 2011, available at http://www.eia.gov/EMEU/cabs/Nigeria/pdf.pdf (accessed 23 February 2012).

12. U.S. Department of Energy, *Country Analysis Briefs: Nigeria*, August 2011, available at http://www.eia.gov/EMEU/cabs/Nigeria/pdf.pdf (accessed 23 February 2012).

13. The quantity ranges for how much oil Nigeria extracts and exports are based on the highest and lowest estimates provided by the NNPC, the OPEC, and the U.S. DOE.

14. U.S. Department of Energy, *U.S. Energy Information Administration: Countries*, 2010, available at http://205.254.135.7/countries/ (accessed 17 July 2012).

15. Sonia Shah, *Crude: The Story of Oil*, New York: Seven Stories Press, 2004, p. 94.

16. For a more thorough-going analysis of the reasons why Nigeria has avoided breaking up, see J.N.C. Hill, *Nigeria since Independence: Forever Fragile?* (Basingstoke, Hampshire and New York: Palgrave Macmillan, 2012).

17. The Economist, *Nowhere Left to Go: Xenophobic Violence Rocks South Africa's Biggest City*, 20 May 2008, accessed 11 December 2010, available at http://www.economist.com/node/11399350

18. The only Nigerians encountered in the film are members of a blood-thirsty gang who make their living through smuggling and profiteering. They are led by a wheelchair-bound sadist who practices *juju* called Obesandjo. The similarity of his name to that of former president Olusegun Obasanjo was not lost on the Nigerian government, which promptly called for the film to be banned.

19. Henry Okah is the leader of the MEND. He was arrested at his home in the early hours of October 2, 2010, at the behest of the Nigerian government following the MEND's bombing of the Ministry of Justice headquarters in Abuja the previous day. Okah had been living in South Africa since his pardon by a Nigerian court the previous July. *The Daily Telegraph*, "Car Bomb Attacks on Nigeria's Independence Day Kill 10," 1 October 2010, accessed 12 October 2010, available at http://www.telegraph.co.uk/news/worldnews/africaandindianocean/nigeria/8037584/Car-bomb-attacks-on-Nigerias-independence-day-kill-10.html p. 1.

20. Daniel Flemes, "Regional Power South Africa: Co-Operative Hegemony Constrained Historical Legacy," *Journal of Contemporary African Studies* April 2009, vol. 27, no. 2, pp. 145 and 155n.

21. The Bakassi Peninsula has a troubled history. At independence, the territory formed part of the new Nigerian state, although Cameroon continued to claim sovereignty over it. Over the years that followed, the dispute became steadily more acrimonious as increasing amounts of oil were discovered in the areas surrounding it. The rancour resulted in frequent armed clashes along the border until Cameroon took the matter to the International Court of Justice (ICJ) on March 29, 1994. On October 10, 2002 the ICJ found in favor of Cameroon. Its decision was supported by the UN, which arranged for final handover talks between Presidents Obasanjo and Biya to be held in June 2006. To win favor with the international community, President Obasanjo agreed to abide by the court's decision. The handover of the territory began shortly afterwards on August 1, 2006 and was completed on August 14, 2008. During that period, the Nigerian Senate tried to stop the process by claiming that the ruling was illegal under Nigerian law. For a brief summary of the conflict see Michael T. Klare, *Blood and Oil: The Dangers and Consequences of America's Growing Petroleum Dependency* (London and New York: Penguin Books, 2004).

22. On January 17, 1983, President Shagari ordered the expulsion of around 3 million foreign nationals from Nigeria. Those instructed to leave were given just fourteen days to do so. The largest group of foreign residents living in the country at that time were Ghanaians. Their government reacted with fury to the order, which violated the May 1979 ECOWAS Protocol on the Free Movement of Persons and Goods. Abergunrin, *Nigerian Foreign Policy Under Military Rule*, pp. 103 and 105.

23. With recent improvements in deep water oil drilling techniques, the oil reserves under the waters of the Gulf of Guinea are being looked at with renewed interest. Yet the maritime boundaries between the various Gulf States have never been well defined. After wrangling over the demarcation of their shared border, the governments of Nigeria and São Tomé e Príncipe agreed, in August 2000, to the establishment of a Joint Development Zone (JDZ). Yet concerns remain in São Tomé over Nigeria's willingness to respect the terms of their agreement. John Ghazvinian, *Untapped: The Scramble for Africa Oil* (Orlando, Florida and London: Harcourt Inc, 2007), pp. 210 and 228–232.

24. Al Jazeera, "Nigeria's Boko Haram Chief 'Killed,'" 31 July 2009, available at http://www.aljazeera.com/news/africa/2009/07/2009730174233896352.html (accessed 14 December 2012). This clash has since become known, both locally and internationally, as the battle of Maiduguri.

25. Vanguard, "Terror: 40 Killed in Christmas Bombings," 26 December 2011, available at http://www.vanguardngr.com/2011/12/terror-40–killed-in-christmas-bombings/ (accessed 14 December 2012).

26. U.S. House of Representatives Committee on Homeland Security Subcommittee on Counterterrorism and Intelligence, *Boko Haram: Emerging Threat to the U.S. Homeland*, 30 November 2011, available at http://homeland.house.gov/hearing/subcommittee-hearing-boko-haram-emerging-threat-us-homeland (accessed 22 May 2012), p. 2.

27. Shell estimates that between 15 and 20 percent of the oil it pumps is stolen. *The Guardian*, "Shell Spends Millions of Dollars on Security in Nigeria, Leaked Data Shows," 19 August 2012, available at http://www.guardian.co.uk/business/2012/aug/19/shell-spending-security-nigeria-leak (accessed 14 December 2012).

28. Marc-Antoine Pérouse de Montclos, "Maritime Piracy in Nigeria: Old Wine in New Bottles?" *Studies in Conflict and Terrorism*, 2012, vol. 35, nos. 7–8, p. 531.

29. *The Guardian*, "Nigeria Braces for Escalation in Terrorist Attacks," 25 September 2012, available at http://www.guardian.co.uk/world/2012/sep/25/nigeria-terrorist-attacks-boko-haram (accessed 14 December 2012).

30. Between 1999 and 2007, the governments of each of the twelve states that make up the region introduced sharia law at the state level. Yet Muslims in the region still get to choose whether they are tried under or have their cases heard in either religious or a secular court, and sharia law does not apply to the North's Christian inhabitants. *Boko Haram* wants to end this and ensure that sharia law is the only legal process in the country and that it is applied to everyone regardless of their religion.

31. Amnesty International, *Nigeria: Trapped in the Cycle of Violence*, 2012, available at http://www.amnesty.org/en/library/asset/AFR44/043/2012/en/04ab8b67–8969–4c86–bdea-0f82059dff28/afr440432012en.pdf (accessed 14 December 2012), p. 9.

32. Human Rights Watch, *Spiralling Violence: Boko Haram Attacks and Security Forces Abuses in Nigeria*, October 2012, available at http://www.hrw.org/sites/default/files/reports/nigeria1012webwcover.pdf p. 40ff.

33. Vanguard, "Dozens of Boko Haram Help Mali's Rebels Seize Gao," 9 April 2012, available at http://www.vanguardngr.com/2012/04/dozens-of-boko-haram-help-malis-rebel-seize-gao/ (accessed 14 December 2012).

34. Human Rights Watch, *Spiralling Violence*, p. 40ff.

35. *The New York Times*, "American Commander Details Al Qaeda's Strength in Mali," 3 December 2012, available at http://www.nytimes.com/2012/12/04/world/africa/top-american-commander-in-africa-warns-of-al-qaeda-influence-in-mali.html?_r=0 (accessed 14 December 2012).

36. Indeed, Mamman Nur, *Boko Haram*'s second-in-command and the mastermind of the attack on the UN, has spent significant time in Sudan.

13 The Republic of Korea

Patrick M. Morgan

THE REPUBLIC OF KOREA (ROK) IS ONE OF THE WORLD'S FOREMOST success stories, the product of a rapid advancement that began about 1961. Before then, burdened by enormous destruction from the Korean War (1950–1953) and political turbulence under President Syngman Rhee, it was not very successful. It made little progress toward political stability—Rhee was autocratic, erratic, and unpopular. It only slowly recovered economically in the 1950s, falling well behind North Korea. Today it is an outstanding and influential example of rapid development and modernization. It views itself, correctly, as a highly advanced economy (approaching the top ten in its GDP), one of the technologically most advanced societies, and rapidly advancing in education, scientific research, information technology, health care, and many other areas, even popular culture. With the world's sixth largest armed forces, it is near the top ten in military expenditures.[1] This is very impressive for a nation of approximately 50 million people. Pride in such accomplishments anchors the national self-image: an advanced nation, rising in stature and influence.

Paradoxically, it is among the world's more insecure states and societies, which deeply affects its foreign relations and domestic politics. The insecurity is multilayered and politically controversial. ROK uneasiness affects its neighbors and closest regional associates; together they constitute an unsettled regional subsystem, periodically caught up in major crises even as outright warfare continues to be avoided. Characterizing this insecurity is a tall order. It starts with being incomplete as a nation. Like China or Pakistan, the ROK lacks a significant portion of what it considers the national territory and population, which is a constant source of tension and confrontation. Part of the insecurity is fear of the missing elements being irretrievably lost. And these are not in a secessionist state but under a rival government, North Korea (DPRK)[2] claiming to be the only legitimate, truly "Korean," government. Insecurity is augmented by the presence of elements in the ROK with some affinity for the North, and concern about provocateurs, intelligence agents, and so forth infiltrated from the North, plus a constant barrage of DPRK propaganda. A contributing factor is that some citizens, regardless of their feelings about the North, think unification must be the paramount objective, outranking the ROK's other goals and concerns.

The most prominent source of insecurity is the military threat posed by North Korea: the world's fourth largest military forces with a fledgling nuclear arsenal, ample stockpiles of chemical and biological weapons, and one of the world's largest special operations forces.[3] Exacerbating this is that the heart of the nation—the greater Seoul area—virtually borders on the DPRK (it is about thirty miles from the Demilitarized Zone), and the bulk of the North's forces, particularly its 800–1,000 missiles and its artillery units, are stationed within 100 kilometers of the border, so the ROK could suffer huge harm in a war even if not invaded.[4] Like Israel, the ROK lacks strategic depth for its most valuable assets; a successful war, even if short, could still be devastating. And North Korea constantly projects intense hostility, including assertions it considers or will consider numerous steps by the ROK and others as "acts of war" that justify a military response. It doesn't help that as an open society the ROK is readily penetrated for political or intelligence purposes, while the DPRK remains closed and opaque.

Beyond this is a broader concern. An old Korean lament about international politics is that "Korea is a shrimp among whales,"[5] at an intersection of the national interests of China, Russia, Japan, and the United States. Each seeks to dominate or potently influence the Korean peninsula because it is too strategically important to cede to the others. For many Koreans, this explains why their nation has been divided since World War 2—Korean insecurity is *geostructural*. Attacks could come in the future from any direction from states with far greater resources. A classic strategy of this "shrimp" was to sharply limit interactions with outsiders—Korea was long known as the "hermit kingdom." North Korea pursues an updated version of it today, but the ROK opted to become one of the world's most interactive societies, with elaborate economic, political, technological, and social links to others.

Allying with the United States is therefore often explained as necessary to offset serious threats in a balance-of-power fashion, despite requiring heavy dependence on an outside power to do so. The United States is the most comfortable ally because it is geographically distant and has no history of wanting to absorb the peninsula. But it can still seem too close because of its forces in Japan and East Asia generally and its unique global power projection. Some Koreans find the United States unnerving as an ally since they believe American national interests will someday shift, ending its commitment to the ROK precisely because of its geographical distance, a concern other American allies share.[6]

A related concern is that perhaps Korea has little intrinsic value to the United States, which cares about it only as an adjunct of its more important commitment to Japan. The idea is that Japan judges the reliability of that alliance in part by the durability of U.S. ties to the ROK; Korea is not important to the United States but it is strategically vital to Japan, which is hugely important to the United States. Being important only derivatively is discomforting.

Next is the insecurity—as in many alliances—from relying on an ally that might act provocatively and involve the ROK in a conflict, even war, not of the South's choosing. The South fears losing control (in alliance theory, the fear of entrapment) over crucial security decisions. Commonly cited is how this could happen in a U.S.-China conflict over Taiwan, with the United States using its forces in Korea in the fighting and insisting on direct ROK military help.

Seoul might have sharply reduced dependence on the United States years ago by using its superior resources to completely outclass the North militarily, particularly after the end of the Cold War left the North with no solid Great Power patron. But getting the North to collapse looked like a grim prospect, and fighting with it would have been very costly. The North is the world's most militarized country—an estimated 25–35 percent of its GDP goes to military matters and some 5 percent of the population is in active military service. Its official policy for years has been "military first."[7]

Trying to outclass the North militarily was inhibited for years by another aspect of depending on the United States. Washington urged Seoul to strengthen its forces but restricted its steps toward military self-sufficiency. The United States sharply discouraged major ROK purchases of modern military equipment from third parties while refusing to sell the ROK arms that they felt were considered too offense-oriented. It restricted ROK arms manufacturing by banning sales of ROK weapons based on American-licensed technologies. (ROK arms transfers have burgeoned recently.) Furthermore, it rejected ROK requests to develop missiles capable of hitting most of North Korea, and it flatly opposed ROK development of nuclear weapons (or other WMD), at one point forcing suspension of early nuclear weapons research. Restricting the ROK in this way was initiated before the Korean War when Washington disliked Seoul's belligerent rhetoric and military provocations vis-à-vis the North, concerns reinforced when the ROK attempted to undermine armistice negotiations to end that war.[8] Only in recent years has the ROK received U.S. approval to develop missiles that can target Pyongyang (three hundred kilometers or more in range) and strong U.S. support for seeking true military self-sufficiency.

Finally, South Korean insecurity is a legacy of the Korean War. Not only did the war level the country, *it was not won*. Unification was not achieved, so the war really never ended. A divided Korea is the most salient vestige of the Cold War. Koreans objected strongly at the time, but the ROK's friends were unwilling to bear the costs and sacrifices involved in achieving unification, signing an armistice that some Koreans have depicted as a betrayal ever since. South Koreans, even if convinced they can win another war on the peninsula, fear the costs and damage would shred their hard-won prosperity. Simultaneously, they fear the possibility of the North collapsing. Noting the huge costs for West Germany of German unification, the South sees the economic, political, and social costs of suddenly governing and integrating the North overwhelming the ROK resources and damaging its way of life, especially since the gaps between the Koreas are much greater than those between the two Germanys.[9]

In summary, the ROK feels it is a prominent nation entitled to be taken seriously, prepared to participate vigorously in world affairs on the basis of its history, achievements, and resources. It wants unification only after North Korea undergoes major changes that erase the conflict and some of the differences between the two—a so-called "soft landing." The result would be a larger nation in population (70 million), economic strength, and potential military might. As for its regional situation, the ROK hopes to help manage security affairs by facilitating bridge-building across various bilateral conflicts (i.e., China–Japan, Russia–Japan) and promoting multilateral cooperation. For instance, early in the Six-Party Talks, Seoul worked closely with the United States and Japan but also with Beijing in seeking to shrink differences among the three great powers. On a larger scale, such activities might sharply reduce the possible threats facing the "shrimp." Meanwhile, it must cope with pressures and frictions of varying intensity that strain relations with its neighbors, hope for the "soft landing" of the North, and promote gradual unification while having to maintain substantial forces in a high state of readiness. These are central elements of its broad security strategy posture.

Nontraditional conceptions of and approaches to security get great attention elsewhere and are certainly appealing to Koreans. The ROK suffered greatly from the East Asian financial crisis in the late 1990s and has serious pollution problems not only from its economic development but from desertification in North China and China's development. Koreans care deeply about human rights, having elected two recent presidents who built political careers promoting democracy and human rights at great personal sacrifice. But traditional security concerns continue to dominate national security policy, and this could well continue for years.

Security Amid Insecurity

There is, of course, a much more favorable side to the security situation. The ROK enjoys being a highly modern, advanced society relative to Russia or China, in some ways on par with Japan, with an attractive and influential culture ranging from its cuisine to the arts and popular entertainment providing a degree of soft power. It basks in intimate association with the world's military behemoth, the hegemonic state in the Northeast and East Asian regional systems. It has huge interactions with the world's three largest economies. Its political relations with all the great powers are amicable, if not always amiable, which is one reason it provided the current UN Secretary General.

Given all this, ROK military security efforts can focus almost entirely on North Korea, primarily the North's forces along the narrow border plus its missile systems. It has not faced a major naval threat since the post–Cold War deterioration of the Soviet navy. Despite living in what is called a dangerous region, it can concentrate heavily on one significant enemy and avoid being overcommitted or stretched too thin across a wide spectrum of threats. Given the current North-South power distribution in conventional forces, being conquered by the North is considered almost impossible. As for participation in regional and global military security efforts the South can, to a considerable extent, determine how far to carry this. Meanwhile, the North Korean nuclear threat has often been considered only modest recently. Many South Koreans think the North will not use nuclear weapons against other Koreans, and Seoul continues to rely on American extended nuclear deterrence even though the nuclear weapons the United States once stockpiled there were removed in 1991. In June 2009, President Obama gave the Seoul government the first *written* U.S. assurance of continued extended—including extended nuclear— deterrence.[10]

However, nasty North-South tensions in 2010–2011 led many to support a possible redeployment of U.S. nuclear weapons, and even more to endorse the ROK developing its own, for bargaining leverage or as the best way to offset the North's.[11] Seoul must also prepare for major shifts in the international environment. Like everyone else, it is coming to terms with the rise of China.[12] It has already adjusted via huge South Korean investments there, making China its leading trade partner (replacing the United States), and by distancing itself somewhat from the Taiwan problem. It assumes major adjustments are coming in relative American and Chinese power in East Asia but wants the United States to remain prominent there and continue offsetting Chinese influence. Some Korean leaders or analysts suggest the ROK be an honest broker to help smooth those changes, while others scoff at the idea.[13]

Another oncoming systemic change is that East Asia or the northern transpacific region will likely become the core area in international politics, displacing the transatlantic region. This has huge *potential* implications. The region's governments are uncertain whether the regional system can evolve so dramatically and yet remain peaceful, and the future of regional system management is unclear. Will it be multilateral in character? Perhaps, but the Asian members lack strong experience with this and remain divided over who should exercise what level of leadership.[14] Will the United States remain prominent or dominant? Perhaps, but many in the ROK expect East Asians to establish greater independence from the United States. Highly cooperative? The region has now had over a generation of peace, facilitating vast progress by many members, but throughout the area many think this cannot continue indefinitely. The basic conception of what the system and its security status will become is not clear, and various governments including the ROK feel they must make at least some preparations for the worst.

Current ROK plans include developing a stronger, very advanced, military capability including a modern blue-water navy alongside its impressive ground and air forces.[15] While national strategy calls for reinforcing growing ROK national power with the U.S. alliance indefinitely, major recent and prospective shifts in the alliance, driven primarily by American plans, raise important questions. Washington's policy is to reinforce the alliance politically while making the ROK rapidly more responsible for its own defense. This includes reducing U.S. forces in Korea while training, equipping, and preparing them for use not just in Korea but almost anywhere in the world. It also includes supporting a major upgrade in ROK military capabilities to boost Seoul's participation in multilateral security management operations almost anywhere, operations frequently U.S. initiated and directed.

Structurally, this means reorienting the alliance from just deterring attacks on Korea and helping *defend it*, if necessary, to tackling broader security concerns and primarily *helping the United States* to resist, and therefore deter, attacks on or violations of the American-fashioned world order. This is reminiscent of the early Cold War when the United States insisted its allies participate in militarily containing the communist bloc almost anywhere if necessary, notably in the Korean and Vietnam Wars, as a contribution to the grand strategy of containment. ROK participation in Vietnam, for example, was extensive and included heavy fighting. But those wars and related activities were seen by many allies as linked to their own security, in that containment was about protecting them from communist expansion in their own region, so participation in containment abroad compensated for direct American involvement in their national defense as part of containment.

Today the projected trade-off is different: for the ROK it consists of gaining more autonomy and responsibility in national defense (with shrinking U.S. participation) in exchange for increased support (political, military) of the United States in security efforts elsewhere, even far from Korea. This is being replicated in other U.S. alliances, particularly NATO and those with Japan and Australia.[16] It is ultimately an attempt to fashion an American-led expeditionary military capability for coping with both international and internal trouble spots, to be used whether there is a broad multilateral sanction for the operations or not.

Many analysts believe the ROK is now militarily more than a match for North Korea, steadily expanding its advantage through its immensely greater resources, more advanced military technology, and more potent friends/allies selling it what it needs. Gradually shifting current U.S. military responsibilities into Korean hands is all that is necessary. As that occurs, the alliance is bound to take on a different character, and perhaps it cannot survive with that. The United States wants the alliance to be a vehicle for attaching the ROK to other democracies (and some other states) in collective system management. The ROK's more recent description of this is the Lee Myung-bak administration's view that Washington and Seoul are fashioning a "strategic alliance" (announced in June 2009).[17]

The allies take pains to indicate this is not directed at China. In fact, Korean officials usually seem relatively unconcerned about a Chinese "threat,"[18] showing more concern about Japan, despite American efforts to promote Japanese–Korean military cooperation.[19] This is somewhat surprising, given Korea's history of troubled relations with China and resistance to Chinese domination. North Korea, long allied or associated with China, is notably uncomfortable about depending on it and tries to sharply limit Beijing's influence on, and even knowledge about, the North. By contrast, the ROK maintains good relations with Beijing for economic reasons, to enhance the North's isolation, and to have Beijing's help on North Korea. A unified Korea might someday develop testy relations with Beijing, but their interactions are so substantial and extensive, so mutually beneficial, that neither will readily renounce them. The only "threat" regularly cited pertaining to China is that it may militarily intervene if the DPRK is collapsing to establish a buffer zone between itself and the vibrant South, block flows of refugees from the North, and protect Chinese investments there. China's apparent diffidence about North Korea's sinking of a ROK naval vessel in 2010 seemed to confirm its deep concern about a possible collapse of the North and was very irritating to Seoul. But no significant shift in relations followed. Thus despite suspicions about its ultimate intentions, China is still not generally seen as hostile or likely to be so, though such a potent neighbor cannot be ignored as a possible threat.

While making plans to join U.S.-led or other multilateral military security operations, the ROK has not yet engaged in them extensively. Seoul was reluctant to send forces to Afghanistan but finally did so from concern about the alliance—analysts in both countries had suggested that failure to do so would be very damaging.[20] The number of troops sent (3,600) was substantial, but they did not take part in combat. ROK naval units have been active in suppressing pirate attacks off Somalia, also a largely noncombat effort. However, the ROK does not restrict its operations in principle like Japan; it accepts possible involvement in fighting. In this regard, ROK special operations forces may ultimately be the most relevant for future missions.

Underappreciated are the complicated implications of the shifting nature of the American alliances. The United States continues to convert its alliances into political and normative communities that include alliances. This went on during the Cold War but, as noted previously, is now going in a new direction, with rising emphasis on collective military effort by communities of common values cooperating for broad security management. In one analyst's summary:

> By structuring an Asian security system that favors democratic forms of social organization and cooperation among liberal states to provide the public goods that undergird the regional order, the U.S. alliance system and the US-ROK alliance within it, the institutional partnership between Washington and Seoul helps Asian states move toward regional community-building based on consensual norms of economic liberalism, good governance, military and diplomatic transparency, and the common security they provide.[21]

This requires adopting a liberal internationalist perspectives in the allies' conceptions of their national interests. The ROK accepts this American-designed and American-led strategic effort. Not clear is exactly what it will mean, other than continuing intense and expensive efforts to make ROK forces even more compatible with U.S. forces for purposes of warfare or other operations almost anywhere. This continues reorienting the elaborate military integration pursued for years for defending against North Korea, an effort that has been incrementally proceeding.

The National Military Posture and Domestic Elements of National Security

The ROK developed significant military forces during the Korean War but fell behind in military power on the peninsula afterward. The North was amply rearmed by its allies as part of its recovery, and it also more rapidly revived economically and moved sooner into rapid industrialization, including arms manufacturing. Thus it held military superiority over the ROK for several decades. The onset of the South Korean "economic miracle" transformed this situation and allowed Seoul to acquire modern, well-equipped forces plus defense industries for producing advanced military equipment.

Today the ROK's large and modernizing forces remain somewhat incomplete. They are undergoing reshaping, retraining, and reequipping in what the United States calls the military transformation. Though well trained, they have had no serious combat experience since Vietnam. They consist primarily of increasingly mobile ground forces, substantial air forces, a growing number of ground-to-ground missiles, and expanding naval forces. The focus has always been on another ground war with North Korea, so South Korea has limited power projection resources, even in its neighborhood.

Its defense sector produces tanks, planes, missiles, warships, and other advanced military equipment, but the ROK is still not self-sufficient and cannot yet handle all its projected military missions and is therefore relying significantly on the United States to produce so-called "bridging capabilities" for the time being, in areas like intelligence or advance communications. It also counts on getting very substantial U.S. contributions in another war. For the time being, American plans still call for rapidly sending huge ground, air, and naval forces in a large war, and for years the allies have conducted massive training exercises to prepare for this influx and its rapid integration into the fighting.

Much like in Pakistan, the emergence of the nation in and through a serious war, plus having a major enemy next door, necessitated a standing military capability at high levels of readiness. As in Pakistan, this eventually resulted in a semiautonomous military sector predisposed to intervene in domestic politics and even seize the government. This reflected, and reinforced, a strong preoccupation with national security *internally*, leading to repression of leftist and other dissident political elements, including those opposing military rule in favor of democracy.

It was under General (later President) Park Chung-hee that Korea's economic miracle was initiated, producing a stunning initial surge into modernity from 1961–1980 and rapid growth since. This was accompanied by escalating public (and American) pressure for more democracy. Remarkably, repressive military rule eventually yielded relatively gracefully to emergence of a vociferous, boisterous democracy by the second half of the 1980s. During this political transformation, the armed forces became more professional and accepted civilian control and a peripheral position in the political system.

Another legacy of the Korean War was an integrated alliance command system; Korean troops served under the UN Command dominated by the United States. After the war the ROK insisted on an alliance and a large American military presence. As in NATO, the presumption that the enemy was primed to attack on short notice led to American insistence on significant integration of the allied forces in weapons, planning, strategy, training, and so forth. In 1978, this led to creating the Combined Forces Command (CFC) with the CFC commander, always an American (once again, like NATO), reporting to two presidents and charged with directing all allied forces in a war. The commander therefore has had considerable responsibility for keeping the forces at high readiness (even though Korean forces are under their national command in peacetime). The justification for the CFC was the classic principle of unity of command, plus anticipation that in a war, integrating the incoming U.S. forces into the fighting would be far easier under a combined command run by Americans. In addition, crucial missions for a war were mostly assigned to the advanced U.S. technological, logistical, intelligence, and communications capabilities, making the choice of an American commander very sensible.

After the Cold War, inter-Korean relations gradually improved and the alliance steadily enhanced its military superiority vis-à-vis North Korea, reducing both fear of a North Korean attack and tolerance of the CFC. Sentiment rose in Korea for the ROK to take full control of its security posture and strategy, including control of operations in a future war. By the early 2000s, the United States was working with the ROK on plans to change the locations of U.S. forces in Korea, cut their numbers, shift more of their responsibilities to ROK forces, and greatly improve both U.S. and ROK military capabilities—the United States pledged $11 billion and the ROK $10 billion in additional military improvements. Allied plans stressed military modernization, including purchases of a great deal of advanced U.S. weaponry and military technology and—as noted previously—further steps to make US and ROK forces more interoperable. Eventually, the Roh, Tae Woo administration adopted the Defense Reform 2020 plan and related measures aimed at making the ROK virtually self-sufficient militarily by 2020, with a total spending target of over $660 billion for, among other things:

- Upgrading existing weapons
- Cutting ROK forces by 26 percent while making them more streamlined and professional and with enhanced mobility, precision strike capacities, improved intelligence, surveillance, reconnaissance, and better command and control
- Continued building of a ROK blue water navy with much broader missions
- Improving ballistic missile defenses[22]

The South Korean push at the highest level for transferring wartime operational command (OPCOM) to the ROK began in 2005, resulting in setting an initial deadline of April 2012 for it to occur. The transfer is still planned but was eventually postponed until the end of 2015 by the two governments.[23] Officially this was due to renewed tensions with North Korea, particularly after the sinking of the ROK warship *Cheonan* (ascribed to a North Korean submarine attack) and financial pressures curbing ROK defense spending. But it was also in response to American complaints that ROK forces would not be ready to take over all the necessary missions, and uneasiness in Korea that the transfer might be a giant step toward ending the U.S. military presence entirely.[24] In addition, the plan calls for shifting from a true unified command to separate commands (one for each side's forces) that ultimately report to a Korean commander while cooperating (in peace and at war) through a network of committees down through the allied forces. This is a major departure from unity of command and may indicate that the United States expects to leave the defense of Korea entirely to the ROK in the foreseeable future.

These developments were preceded, and enabled, by the political changes involved with the onset of democracy in the late 1980s. As military rule faded, previously repressed elements were soon contending for important political influence and positions, including the presidency. Long-time dissident and champion of democracy Kim, Dae-jung was elected president in 1997, ushering in a decade of efforts to fundamentally reorient national security policy. In contemporary terminology, the government sought to apply nontraditional approaches to security to the problem with the North and to the reordering of Northeast Asian security.

President Kim (1998–2002) moved to strengthen ROK relations with great powers active in Northeast Asia simultaneously, as political groundwork for a major overture to Pyongyang. His new Sunshine Policy attempted to normalize relations between North and South and induce other governments to cancel efforts to isolate North Korea—a strategy based on West Germany's effort in the 1970s to reach out to East Germany. The goal was a less belligerent, less isolated North Korea, better off economically and more in touch with the world. It was no longer depicted as an arch enemy, but as a regime governing suffering Korean compatriots, less an enemy than a state and society needing help. Aid to and trade with the North was significantly expanded, and a summit meeting of the two heads of state was arranged. Cross border contacts improved somewhat, and the ROK government facilitated significant South Korean investments in the North.[25]

President Kim's seizing the initiative in this fashion naturally stimulated more discussion in the ROK about adopting an even more independent foreign policy, which was coupled with rising expressions of anti-Americanism. But the Clinton administration's drive for engagement with Pyongyang meant that the allies' policies on North Korea were basically aligned. When President Roh, Moo-hyun (2003–2007) took over, the strategy for improving North-South relations added a significant additional component through less emphasis on U.S.–ROK ties alongside further expansion of North-South economic and other interactions. The president and his supporters began suggesting that the *American threat to the North* was the primary reason for DPRK suspicion and belligerence, inhibiting better inter-Korean relations. The North was not about to attack the South. The real threat was that hard-line U.S. policies (under the Bush administration) would provoke a war with the North and drag the South into it. U.S. and ROK national security policies were now sharply at odds, with the United States seen as something of a threat by its ally!

Even for those taking ROK complaints and assertiveness less far, it became common to assert that the alliance needed substantial alterations, that it was "unbalanced." It was still a *patron-client* alliance, reflecting the pronounced inferiority and dependence of South Korea, but the ROK was now far more self-sufficient and deserving of more respect and responsibility. The alliance had to be significantly "rebalanced," reducing American dominance in favor of a mature, more equal partnership.

U.S. political responses, particularly in Congress, were initially far from polite, when expressions of ROK anti-American sentiment in opinion polls and media rose sharply, especially in the younger generation, which is some 40 percent of South Korea's population. There were serious concerns the alliance might collapse, due to radical steps by Seoul or rising sentiment in Washington to give up on Korea in disgust. Many Koreans described the alliance as still reflecting American priorities and interests that were classically realist and thus very parochial: retaining U.S. military and political dominance in the East, particularly Northeast, Asia, while continuing to contain its allies and keep them in line with U.S. policies, thus reinforcing U.S. hegemony. Many American analysts traced this to a generational social-political shift, citing the rise of new ROK voters with no personal experience of the huge U.S. assistance during and after the Korean War plus the baleful effect of leftist elements in Korean educational systems on younger citizens' views of the United States.[26]

The Korean armed forces might have been expected to protest strongly against this challenge to the alliance, emerging as it did from a political sector that had been strongly antimilitary, but that response was muted somewhat by the Roh government's commitment to heavy spending for military modernization and its determination to get wartime control of allied forces turned over to Korean officers.

The Bush administration was highly critical of such South Korean engagement policies. However, its relatively quick response to the ROK military plans was to endorse them! The United States said, in effect: "You want fewer U.S. troops in the country? Fine! Command in the next war? Sure! Take over responsibility for your national defense? Good!" This really should not have been a surprise. For several decades presidents had proposed cutting U.S. forces in Korea, starting with President Nixon, and after the Cold War the United States again indicated it wanted to cut and adjust those forces. Now it announced plans to cut its forces by a third, to move the rest south of Seoul and the Demilitarized Zone, and equip and train ROK forces for taking over key wartime U.S. missions. It proposed an early deadline (2012) for a CFC transfer. As noted previously, its emphasis in the alliance shifted toward arrangements for the ROK to join U.S. efforts in global and regional security management.

Pushing on this open door soon had some Koreans suspecting they might not want to go through with it after all, at least not quickly. The American reaction suggested that a classic realist conception was missing something, that a liberal internationalist perspective was now helping shape how American leaders define national interests. Accepting the entire U.S. package seemed to involve too many gambles and uncertainties. Opposition to dissolving the CFC in 2012 rose, citing various concerns:

- ROK unreadiness
- The costs of replacing the departing U.S. forces
- The message such changes in U.S. forces would send to the North
- Pyongyang's heightened belligerence once President Lee Myung-bok and then President Obama were in power

Much has shifted in recent years. The North proved strongly resistant to the ROK engagement strategy in the end—after all, its leaders could see how it had turned out for East Germany. They rejected reforming their regime out of existence—ROK concessions, incentives, and aid brought little in the way of reform, concessions, openness, and the like, and nothing halted the North's nuclear weapons program. The Lee, Myung-bak administration (2008–2012) therefore abandoned conciliation for a hard-line policy of "conditional engagement"—either major reciprocity from the North or no more assistance. North-South relations deteriorated sharply, even though President Lee had proposed a "grand bargain" whereby a DPRK retreat from nuclear weapons would be compensated by ROK efforts to boost North Korea's GDP per capita annual income to at least $3,000. The Obama administration efforts to be conciliatory were equally unsuccessful, leading to a hard line also as the administration emphasized repairing the alliance by closely aligning U.S. and ROK policies. The U.S. position on further negotiations with the North on its nuclear weapons program hardened, delaying resumption of the Six-Party Talks. U.S., ROK, and Japanese pressure on the North handed Beijing a much heavier burden in supporting it. Greater sanctions were imposed, along with tougher efforts to interdict North Korean military shipments to reduce its hard currency income from selling weapons and related technologies. Even humanitarian aid was cut.

The rise in tension was obvious. As noted earlier, it culminated in the sinking of a ROK ship (the *Cheonan*) and an exchange of shelling, with fifty ROK citizens and military personnel killed, followed by two allied naval maneuvers staged off North Korea, involving a U.S. carrier, to show how serious things had become.[27] But repairing the alliance and the delay in shifting wartime control did not alter the basic plan: shift more responsibilities to the ROK within the framework of the alliance while integrating it into American grand strategy globally. The alliance was therefore in better shape than before, with much of the earlier friction eased or set aside, and with the ROK more heavily focused on North Korea's military threat than it had been in over a decade.

North Korean leader Kim, Jong-il's sudden death in late 2011 provoked speculation about a possible thaw in North-South and DPRK-U.S. relations but not much optimism. No major DPRK domestic reforms followed, there were no rapprochements, no renewed contacts and negotiations. The new ruler, the very young Kim, Jong-un (Kim, Jong-il's youngest son), announced that the military first policy would continue, and the regime made two provocative attempts to launch a missile into space, the second one successfully. Optimists had to be content with signs the new regime and its leader might be more open and were serious about trying to improve the welfare of their citizens, plus the fact that the North has often been unusually confrontational as a prelude to reentering negotiations. With no high expectations that negotiations will generate a breakthrough, they may well be undertaken anyway because the North Korean problem has to be at least managed—it never dies down by itself.

Reflections and Speculation

In the coming decades the ROK must cope with a series of dilemmas in its national security efforts. It is caught between the desire to consume security within the world order that the United States mainly provides and the urge to help operate that order. It is bound to have considerable fear of entrapment in pursuing the latter despite its desire to be a major player in regional and global security management. It will also be caught between the burdens of becoming a major player and natural pressures to curb military spending and avoid military entanglements in far-off places.

The profound necessity of adjusting to the rise of China is complicated by uneasiness as to where China is going domestically, regionally, and globally. Adjusting to China means thinking hard about the US-ROK relationship, but here, too, there is real uncertainly as to where the United States is going and what it will do. Is it best to align Korean security with East Asia's emergence as the main focal point of world politics and Korea's place in that development? Or to align it with the West's current hegemony and its efforts to instill its perspectives and values globally, as the ultimate in security in an evolving international politics regardless of where its "center" ends up?

The ROK must also cope with continuing tension between focusing above all on the *North Korean problem* as its foremost national security matter and the key to its security, or setting that problem aside to pursue a larger role in the world. The latter means letting that problem burn itself out and evading the potentially serious costs of trying to continue deliberately resolving it relatively soon. But that would mean largely ignoring one of the world's most vicious regimes and its WMD arsenal right next door, a regime that takes very badly to being ignored.

Notes

1. See the following: *The Military Balance 2011* (London: International Institute for Strategic Studies, 2011), pp. 208–209, 251–254.

2. The Democratic People's Republic of Korea.

3. *The Military Balance 2011*, pp. 205–207; 249–251; Peter Hayes, "North-South Korean Elements of National Power," April 27, 2011 at http://www.nautilus.org/publications/essays/napsnet/reports/north-south-hayes.

4. Bruce E. Bechtol, Jr. "Preparing for Future Threats and Regional Challenges: The ROK-U.S. Military Alliance in 2008–09," in *Shifting Strategic and Political Relations with the Koreas*, Joint U.S.-Korea Academic Studies, vol. 19 (Washington, D.C.: Korea Economic Institute, 2009), pp. 75–99; Stanley B. Weeks, "Coordinating U.S.-ROK Defenses against North Korean Nuclear/Missile Threat," in Jung-ho Bae and Abraham Denmark, eds., *The U.S.-ROK Alliance in the 21st Century* (Seoul: Korea Institute for National Unification, 2009), pp. 221–244.

5. For example, see Ralph A. Cossa, "Needed: A Joint ROK-U.S. Strategy for Dealing with North Korea," in Bae and Denmark, eds., *The U.S.-ROK Alliance in the 21st Century*, p. 280.

6. See Joo-Hong Nam, *America's Commitment to South Korea: The First Decade of the Nixon Doctrine* (Cambridge: Cambridge University Press, 1986), pp. 175–180.

7. *The Military Balance 2011*, 205–207.

8. See Bae-ho Hahn, "Major Issues in the American-Korean Alliance," in Yongnok Koo and Dae-sook Suh, eds., *Korea and the United States: A Century of Cooperation* (Honolulu: University of Hawaii Press, 1984), pp. 91–110.

9. While there are grounds for thinking the DPRK could soon disintegrate, most analysts are very pessimistic about this happening.

10. Richard Fontaine and Micah Springut, "Coordinating North Korea Policy: An American View," in Bae and Denmark, eds., *The U.S.-ROK Alliance in the 21st Century*, p. 151.

11. Ralph Cossa, "U.S. Nuclear Weapons to South Korea?" July 26, 2011. Available at the PacNet Newsletter—PacNet@hawaiibiz.rr.com.

12. See Changhee Park, "Why China Attacks: China's Geostrategic Vulnerability and Its Military Intervention," *Korean Journal of Defense Analysis*, 20, no. 3 (2008), pp. 263–282.

13. James Przystup, "The U.S.-Japan Alliance and the U.S.-ROK Alliance," *Shifting Strategic and Political Relations with the Koreas*, Joint U.S. Korea Academic Studies, vol. 19 (Washington, D.C.: Korea Economic Institute, 2009) pp. 66–67.

14. Sample discussions on this: Niklas Swanstrom, "Security and Conflict Management in East Asia," *The Korean Journal of Defense Analysis*, 20, no. 3 (2008), pp. 187–198; Kent Calder, "Northeast Asia: The 'Organization Gap' and Beyond," in Lee-Jay Cho, Chung-si Ahn, and Choong-nam Kim, eds., *A Changing Korea in Regional and Global Contexts* (Honolulu and Seoul: East-West Center and Seoul National University Press, 2004), pp. 27–74; eight essays on "Security in East Asia: The Pieces of a New Architecture" in *Global Asia* 5 no. 1 (Spring 2010), pp. 8–48.

15. Plans were announced by President Kim back in 2001 for a "strategic mobile fleet." See Weeks, "Coordinating U.S.-ROK Defenses Against North Korean Nuclear/Missile Threat" in Bae and Denmark, *The U.S.-ROK Alliance in the 21st Century*, p. 234.

16. See Sheldon W. Simon, "Theater Security Cooperation in the U.S. Pacific Command: An Assessment and Projection," *NBR Analysis* 14, no. 2 (August 2003).

17. The alliance adjustment was first announced by Presidents Bush and Lee in 2008, then in the statement "Joint vision for the Alliance of the United States of

America and the Republic of Korea" on June 16, 2009. Sung-han Kim, "Strengthening of the ROK-US Alliance for the 21st Century" and Weeks, "Coordinating U.S.-ROK Defenses Against North Korean Nuclear/Missile Threat," in Bae and Denmark, *The U.S.-ROK Alliance in the 21st Century*, pp. 221–244 and 345–370.

18. Chang Jae Ho, "The 'Rise' of China and its Impact on South Korea's Strategic Soul-Searching," in *The Newly Emerging Asian Order and the Korean Peninsula*, Joint U.S.-Korea Academic Studies, vol. 15 (Washington, DC: Korea Economic Institute), 2005, pp. 1–11; Robert Sutter, "The Rise of China and South Korea," *The Newly Emerging Asian Order and the Korean Peninsula*, pp. 13–34; an example of greater concern about China: Taeho Kim, "Sino-ROK Relations at a Crossroads: Looming Tensions amid Growing Interdependence," *The Korean Journal of Defense Analysis* 17, no. 1 (2005), pp. 129–149.

19. See Chung-in Moon and Chun-fu Li, "Reactive Nationalism and South Korea's Foreign Policy on China and Japan: A Comparative Analysis," *Pacific Focus* 25, no. 3 (2010), pp. 331–355; Cheol-hee Park, "The Future of Korea-Japan's Strategic Relationship: A Case for Cautious Optimism," *Shifting Strategic and Diplomatic Relations with the Koreas*, pp. 101–118.

20. David Straub, "U.S. and ROK Strategic Doctrines and the U.S-ROK Alliance," in *Dynamic Forces on the Korean Peninsula: Strategic and Economic Implications*, Joint U.S.–Korea Academic Studies 17 (2007), p. 180.

21. Daniel Twining, "Strengthening the U.S.-Korea Alliance for the 21st Century—The Role of Korean-American Partnership in Shaping Asia's Emerging Order," in Bae and Denmark, *The U.S.-ROK Alliance in the 21st Century*, pp. 333–334.

22. Przystup, "The U.S.-Japan Alliance and the U.S.-ROK Alliance," *Shifting Strategic and Political Relations with the Koreas*, pp. 63–65.

23. Ibid.; Bechtol, *The Military Balance 2011*, p. 209.

24. Ralph A. Cossa, "Needed: A Joint ROK-U.S. Strategy for Dealing with North Korea," in Bae and Denmark, *The U.S.-ROK Alliance in the 21st Century*, pp. 275–316.

25. On the Sunshine Policy see Ralph Cossa and Alan Oxley, "The U.S.-Korea Alliance," in Robert D. Blackwill and Paul Dibb, eds., *America's Asian Alliances* (Cambridge, Mass.: MIT Press, 2000), pp. 61–86.

26. The US-ROK dispute is examined in Choi Kang, "A View on America's Role in Asia and the Future of the ROK-US Alliance," in *The Newly Emerging Asian Order and the Korean Peninsula*, pp. 217–233; Sook-jong Lee, "Allying with the United States: Changing South Korean Attitudes," *The Korean Journal of Defense Analysis* 17, no. 1 (Spring 2005), pp. 81–104; Gi-wook Shin, *One Alliance, Two Lenses: U.S.–Korea Relations in a New Era* (Stanford, Calif.: Stanford University Press, 2010).

27. The National Institute for Defense Studies, Japan, *East Asian Strategic Review 2011* (Tokyo, *The Japan Times*, 2011), pp. 3–4, 89–91; Charles A. Armstrong, "South Korea and the Six-Party Talks: The Least Bad Option?" in *Tomorrow's Northeast Asia: Prospects for Emerging East Asian Cooperation and Implications for the United States*," in Joint U.S.–Korea Academic Studies, (Washington, D.C.: Korea Economic Institute, 2011), pp. 165–177.

14 Turkey's New (De)Security Policy: Axis Shift, Gaullism, or Learning Process?

Bill Park

The End of the Cold War and the Beginnings of Regional Engagement

The Cold War's demise offered Turkey unprecedented opportunities to access the peoples and states of the various regions with which the country overlaps. The former Ottoman domains in the Balkans, Black Sea, and Caucasus regions now opened up, and the outbreak of multiple conflicts in these regions served to impel as well as attract Turkish engagement. Turks were particularly keen to reconnect with their long-forgotten ethnic Turkic cousins in Azerbaijan and in central Asia. The collapse of the Soviet Union was followed by the development of relationships with Russia, Ukraine, and with the rest of post-communist Europe which, like Turkey, now aspired to join the EU. Turgut Ozal, first as prime minister and then as president, was closely associated with this new activism in Turkish foreign policy.[1] Ismael Cem, foreign minister from 1997 until 2002, similarly adopted a more engaged diplomatic approach than had hitherto been associated with Turkey, notably toward Greece and Syria. During the 1980s Ozal had also engineered major reforms in the Turkish economy, laying the foundations for Turkey's more recent emergence as a major trading state—and for its increased dependence on imported energy, largely from Russia and Iran.[2]

A distinctive and, for Turkey, somewhat novel characteristic of this more active regional engagement were the attempts to desecuritize some of Turkey's hitherto more fraught relationships, most particularly that with Greece.

Foreign Minister Cem successfully introduced low politics confidence-building measures into this usually fraught relationship. This entente has endured, and the two neighbors have achieved closer cooperation on tourism, environmental issues, travel between the two countries, and on the handling of illegal immigration across their shared borders. Their economies are more interlocked, and military confidence-building measures have been agreed on.[3] Greece now supports Turkey's EU accession, and the two countries seek to avoid direct conflict over the Cyprus issue. In 2005 they established a hot line between their two air forces as a crisis management measure in the face of the continued mutual allegations of and tensions over airspace violation—although such incidents occur to this day.

Another bilateral relationship that underwent a remarkable transformation was that with Syria. This relationship had long been troubled by Syria's unease concerning the implications for the water levels of the Tigris and Euphrates rivers, which rise in Anatolia and flow into Syria and Iraq, of south-eastern Turkey's vast irrigation and hydroelectric power project. Another bone of contention was Turkey's 1939 incorporation of the Hatay province, which Damascus continued to regard as rightfully Syrian. The 1998 threat by Turkey of military action against Syria because of its support for the PKK resulted in the immediate expulsion by Damascus of PKK leader Abdullah Ocalan and the closure of PKK bases in Syrian-controlled Lebanon. In the immediate aftermath of the crisis, the two neighbors embarked on a significant expansion of economic, political, security, and cultural links. The water and Hatay issues were put on the diplomatic backburner.[4]

Turkey also embarked on multilateral diplomatic initiatives aimed at regional desecuritization, for example with the establishment in 1992 of the Black Sea Economic Cooperation (BSEC) at Ankara's initiative. BSEC now has twelve members and is dedicated to enhanced cooperation in low politics confidence-building areas such as transport, combating crime, tourism, environmental protection, and trade and economic development. Turkey also took the lead in establishing the Black Sea Naval Cooperation Task Group (BLACKSEAFOR) in 2001, along with Russia, Ukraine, Georgia, Bulgaria, and Romania, which aims to contribute to security, stability, and maritime cooperation in the area. Although the fruits of such endeavors have been limited, they nevertheless offered early indications of a relatively unfamiliar soft power cooperative approach to security on Ankara's part.

During the 1990s a succession of fractious coalition governments nevertheless helped ensure that any shifts in Turkey's foreign and security policy approach remained piecemeal. Security policy remained primarily in the hands of the country's Kemalist military and state bureaucracies. There were few signs of fresh thinking as such, and a "hard" securitized approach to foreign policy remained a compelling feature of Turkey's behavior. Thus, throughout the 1990s Turkey conducted frequent cross-border raids into northern Iraq in pursuit of the PKK. In 1993 Turkey closed its border with Armenia as a result of that country's conflict with Azerbaijan over the enclave of Nagorno Karabahk (it remains closed to this day). In 1996 it came close to blows with Greece over an uninhabited island in the Aegean (and Turkish-Greek differences over the Aegean remain unresolved). In 1998, the same year Ankara built up forces on its Syrian border, it threatened a military response should Greek Cyprus agree to the deployment of Russian S-300 antiaircraft missiles on Cyprus. Indeed, Ankara exhibited little flexibility regarding Cyprus and retained (and still retains) substantial forces in the island's north.

The Emergence of the AKP and the Invasion of Iraq; Impulses for Change

The AKP's 2002, 2007, and 2011 election victories have endowed Turkey with single-party governments for over a decade now, offering scope for a more single-minded and consistent approach to foreign and security policy. Within months of its November 2002 victory, the U.S. spearheaded an invasion of Iraq, which in its now infamous March 2003 vote the Turkish parliament had refused to facilitate.[5] These two developments brought about a profound shift in Turkey's approach to its external relationships. The vote faithfully reflected Turkish political, bureaucratic, and public sentiment across the political spectrum. The U.S. presence in Iraq and the close relationship Washington forged with the Kurdish Regional Government (KRG) of the north served to worsen still further the U.S.–Turkish relationship. Ankara now could not and the United States and the KRG would not tackle PKK forces based inside Iraq. Turks suspected KRG collusion in and U.S. indifference to the increase in cross-border PKK attacks that followed. Anti-American sentiment in Turkey escalated dramatically. Turkey's mounting frustration eventually forced Washington's hand, and in November 2007 President George W. Bush agreed to provide Turkey with "actionable intelligence" on PKK movements and bases and gave the green light to Turkish military incursions across the border into northern Iraq. Turkish bombing raids and ground incursions into the KRG zone—and PKK attacks inside Turkey of course—have continued ever since.

The transborder Kurdish issue was seen in Turkey as an existential security threat, while the chaos in Iraq raised the specter of a wider regional destabilization. In the wake of the U.S.-led attack on Iraq in 2003, Ankara found that its immediate Arab and Iranian neighbors were more in tune with its concerns than was its U.S. ally. Accordingly, the then foreign minister Abdullah Gul secured agreement for a regional summit to be held in Istanbul in January 2003, attended by Egypt, Syria, Jordan, Saudi Arabia, and Iran. This loose regional alliance has subsequently become known as the Platform for Iraqi Neighbors. Joined in 2004 by the new Iraqi government, around a dozen meetings of the Platform have since been held. For Turkey, forays such as this into Middle Eastern politics had hitherto been unusual. The Middle East has since emerged as a major priority for Turkish policy-makers.

The Kurdish issue, and the related Turkish concern over Iraq's future more broadly, provided reasons for Ankara's particularly assiduous cultivation of Syria and Iran. Ankara arrived at close intelligence and operational understandings with Tehran and Damascus regarding the activities of Kurdish nationalists. With respect to Iraq's political evolution, Ankara wooed all the main factions there, was instrumental in persuading the Sunni leaderships of the wisdom of participating in the 2005 elections, and mediated between Iraq's Sunni leaderships and the United States and between Iraq and Syria.[6] In fact, Ankara's objectives increasingly overlapped with Washington's. Both sought the emergence of a politically viable Iraq and its reincorporation into a stable regional order to minimize the scope for and temptation of Kurdish independence and to counter Iranian influence in the country.

Davutoglu's Philosophy

However, it is the emergence of the figure of Ahmet Davutoglu, first as foreign policy advisor to Prime Minister Erdogan and then, from 1 May 2009, as foreign minister, that has given voice, direction, and drive to Turkey's new foreign and security policy. Representing a political party that is at odds with many aspects of Turkey's Kemalist legacy, Davutoglu has laid out his ideas in his book *Strategic Depth*[7] and in numerous speeches, interviews, and journal articles.[8] The geopolitical concept of strategic depth constitutes his core guiding principle. For Davutoglu, Turkey is neither a bridge between two points nor is it peripheral to any region. Rather, it lies at the geopolitical and geostrategic heart of a vast region incorporating the Middle East, the Balkans, the Caucasus, Central Asia, the Caspian, the Mediterranean, the Gulf, and the Black Sea. These regions are linked by Turkey and in turn constitute Turkey's hinterland. This geography both enables and obliges Turkey to adopt an active diplomatic agenda, as its own security and economic well-being depends on a wider regional tranquillity. This multidirectional aspect to Turkish policy is captured by the phrase "zero problems with the neighbors." This requires more than an improvement in Turkey's bilateral relationships, although these have certainly received Ankara's energetic attention. It has also been expressed through a commitment to dialogue, engagement, multilateralism, confidence-building measures, dispute mediation, trade agreements, the institutionalization of diplomatic relationships, economic aid and reconstruction, and peace support operations.

Davutoglu frequently stresses Turkey's historical and cultural connections with much of its neighborhood, not least the shared Ottoman past. This partly explains why Turkey's new foreign policy is also sometimes dubbed "neo-Ottoman."[9] However, the AKP's regional foreign policy has not been confined to former Ottoman territories. Nor indeed does it exclusively favor Turkey's Islamic partners. Thus, Ankara has also sought to establish constructive relations with its former imperial adversary Russia, with "Christian" Georgia, Greece, Croatia, Serbia, Ukraine, Romania, and Bulgaria, and with central Asia, which never constituted part of the Ottoman Empire.

Davutoglu also foresees a global shift in the balance of power, away from the West and toward Eurasia and the developing world more generally. He believes that the world is moving toward a multipolar, interdependent, less Western centric and more diverse, inclusive and participatory order. He associates Turkey with this global shift and with the so called BRIC countries (Brazil, Russia, India, and China) and with other rising powers such as South Korea. This identification with the world's emerging powers is not an entirely fanciful notion in the light of Turkey's impressive economic growth levels in recent years. In its immediate region, Turkey has the second largest economy after Russia's, and the seventeenth in the world. As a member of G20, Turkey regards this gathering as more representative of the new world order than the more exclusive G8 and as an instrument to protect and further the interests of the developing world against the economic haves of the West. Turkey has also been busily opening embassies and dispatching trade delegations in regions as distant from Turkey as sub-Saharan Africa and Latin America. These strategies entail expanded participation in regional trade fairs, government-sponsored campaigns, trade delegations, ministerial visits, the establishment of direct flights by Turkey's national air carrier, Turkish Airlines, and so on.

"Soft" Power as an Instrument of Turkish Security Policy

Indeed, Turkey's more active and geographically diversified external policies should in part be seen as driven by the country's emergence as a trading state.[10] The rise, since Ozal's reforms in the 1980s, of export-oriented businesses in Turkey's pious heartlands has been instrumental in developing Turkish trade relations in Middle Eastern, former Soviet, and African markets. Trade is thus significantly more diversified than was formerly the case. Turkey's regional initiatives are frequently closely associated with its pursuit of its expanding economic interests. Davutoglu invariably travels with an entourage of Turkish businessmen and hopes that trade with Turkey's neighbors will interlock their fates with its own. Turkey's economic growth has also brought with it an increased dependence on imported energy, notably from Russia and Iran. Added to this, Turkey's ambition to be an energy transit route linking Middle Eastern and Caspian basin reserves to markets in the west, bypassing Russia, and also to be an energy hub in its own right, further underpin much of Ankara's energetic diplomacy.[11]

Manifestations of Davutoglu's desecuritizing approach to diplomacy abound. For example, Turkish initiative has produced visa-free regimes with a range of nearby countries including Syria, Russia, Lebanon, Libya, and the Gulf states, and bilateral high-level strategic cooperation councils with countries such as Syria, Iraq, Azerbaijan, Kyrgyzstan, and a host of others. A multilateral council incorporating Turkey, Syria, Lebanon, and Jordan has also been established. In this context, Ankara's aborted attempt to normalize relations with Armenia deserves a mention.[12] Davutoglu's numerous attempts at mediation have produced outcomes such as the Istanbul Declaration of April 2010 with Turkey, Serbia, and Bosnia, whereby Serbia recognized Bosnia's territorial integrity and sovereignty. An initiative with Turkey, Croatia, and Bosnia is focusing on infrastructure projects and reconstruction in the region.

Turkey has also actively sought reconciliation between Serbia and Kosovo on the basis that only this will produce a lasting peace in the region and ensure Kosovo's security. Turkey has hosted annual meetings between the Afghan and Pakistani heads of state since 2007 and has brought together intelligence and military officials. In January 2010, it also brought them together with neighboring and other interested countries such as China, Iran, the central Asian states, and the UK. Ankara's inclinations are toward "Afghanization" and regionalization of the approach to that country's travails. It has sought to distance itself from Washington's war with the Taliban, who Ankara believes should be brought into Afghanistan's political process. Israeli-Syrian, Syrian-Iraqi, and Iranian-U.S. relations have also been targets of Turkish offers of mediation, as have intrastate and inter-communal disputes between Iraqi, Palestinian, Lebanese, and more recently Libyan factions.

Perhaps the most remarkable expression of Turkey's new approach to security issues relates to its new-found friendship with the KRG in northern Iraq.[13] In May 2008, and in stark contrast to Turkey's position hitherto of refusing to deal directly with the KRG, Davutoglu embarked on a series of meetings with the KRG leadership. Close economic, political, and security cooperation between Ankara and Erbil has since developed, and in 2010 Ankara opened a consulate in Erbil. Northern Iraq has already emerged as a major market for Turkish goods and construction companies, and in 2009 Turkey began directly importing oil from northern Iraq. Indeed, Erbil's need for Turkey as an outlet for its energy exports, and Turkey's energy needs and aspirations, constitute in themselves a rationale for the burgeoning relationship between them. In short, Ankara's new approach to security has led it to embrace rather than confront the Kurds of northern Iraq.

Initiatives such as these have emerged as central tools of Ankara's approach, which emphasizes soft rather than hard power.[14] Indeed, should the AKP government's approach to security problems meet with long-term success, it might conceivably lead to a weakening of Turkey's inflated military establishment. This prospect could be reinforced by the domestic political struggle currently in train over the political role of the general staff. This is largely focused on the so-called Ergenekon investigation into the activities of the deep state, which has led to the arrest and detention of hundreds of retired and active military officers[15] but which is also driven by the EU accession requirement for civilian control over the military, by Turkey's ongoing debate about a new constitution, and by the AKP's more general challenge to the country's Kemalist establishment and legacy. Given the role the military has traditionally played in setting the agenda of Turkey's security policy, and indeed of securitizing it, this could in turn lead to the emergence of a dynamic in internal Turkish politics that could reinforce current foreign policy trends. On the other hand, the prospect of a hard-line reaction to Turkey's current direction cannot be ruled out.

Axis Shift, or Turkish Gaullism?

The regionalized, "third worldist," neo-Ottoman, and perhaps even Islamized tone to Turkey's recent foreign policy has sometimes irritated Ankara's traditional Western allies. To be fair, these raised eyebrows have often been prompted more by Prime Minister Tayyip Erdogan's emotive statements than by Davutoglu's actions, but Turkey's recent external policy approach can readily be interpreted as a paradigm or axis shift.[16] Among Erdogan's contributions can be listed his numerous emotive remarks relating to Israel, his declaration of friendship with Iran's President Ahmedinijad in the wake of the latter's questionable election victory, his November 2010 receipt of Gaddafi's International Prize for Human Rights, his query as to what interest NATO could possibly have in Libya on the eve of its intervention there, his strident opposition to the 2009 appointment of Andres Fogh Rasmussen as NATO Secretary General of NATO, his defense of Sudanese President Omar al-Bashir against his indictment for war crimes and his associated declarations that "a Muslim can never commit genocide" and that in any case Israel's crimes were worse.

More substantive has been Ankara's dramatic fallout with Israel. Davutoglu's attempt at mediation between Israel and Syria collapsed as a consequence of Israel's attacks on Gaza in late 2008. Turkey's relations with Israel plunged still further when in January 2009 Prime Minister Erdogan stormed from the World Economic Forum platform he was sharing with Israeli President Shimon Peres. In October 2009, Turkey cancelled Israeli participation in an air force military exercise to be held in Turkish airspace, leading Italy and the United States to withdraw in protest. More dramatically still, at the end of May 2010 a convoy of ships laden with humanitarian aid left Turkish ports with the intention of defying Israel's Gaza blockade. An Israeli commando raid on the flotilla resulted in the loss of eight Turkish and one Turkish-American life. Turkey withdrew its ambassador to Israel and relations plunged still further.[17] Yet, until this sudden deterioration, trade, tourism, and government-to-government cooperation had been growing impressively. In November 2007, President Shimon Peres became the first Israeli leader to address the TGNA. Furthermore, in the wake of the tiff Turkey kept bilateral contacts open and expressed its wish for an eventual restoration of the relationship. In any case, Turkish public opinion was sympathetic to the Palestinian plight long before the AKP came to power, and to some degree the government's position can be seen as one that is representative of the electorate on which it relies for its domestic support.

Another substantive cause of concern to the West has been Ankara's approach to relations with Iran, which under the AKP have amply demonstrated Ankara's general preference for diplomatic engagement rather than confrontation.[18] The energy relationship between the two countries had already incurred Washington's displeasure,[19] and Turkey's approach to the Iranian nuclear issue served only to intensify it. Ankara insists that Tehran should comply with the resolutions of the International Atomic Energy Agency (IAEA) and permit inspection of its nuclear facilities. However, Ankara also opposes either military action or the use of further sanctions against Iran.[20] Thus, it abstained in the IAEA's November 2009 resolution that condemned Iran's evasiveness even though Russia, China, and India voted for it. Matters appeared to deteriorate further when Turkey and fellow UNSC member Brazil negotiated an agreement involving the depositing of some of Iran's low-enriched uranium in Turkey just as Washington's intensified diplomatic efforts to persuade the UNSC to adopt still tighter sanctions against Iran seemed to have met with the approval of all five permanent members. Amidst fierce U.S. anger, Ankara then voted against the UNSC resolution aimed at tightening sanctions on Iran.[21]

Yet it has also been plausibly argued that Turkey has not so much shifted its axis as declared its independence, and that Ankara's foreign and security policies have acquired a Gaullist tinge.[22] As a growing and increasingly self-confident country, well positioned to exploit the greater global diplomatic fluidity that now exists, Ankara's traditional allies might arguably be neither surprised nor disturbed by its novel diplomatic activities. One way in which this independence is being expressed is in the military procurement sector. Not only is Turkey steadily expanding its self-reliance in military production, but it is also prepared to look further afield in its search for collaborative partners. Russia, China, and South Korea are increasingly competing with the United States, Europe, and Israel for Turkish military business.

Turkey as an Ally, as a "Bridge," and as a "Model"

Neither the axis shift nor the Gaullist interpretations of Turkish policy offer the full picture. Since the Cold War's end, in the wake of the 9/11 attack, and continuing under the AKP government, Turkey has contributed to multilateral initiatives and operations alongside its Western allies and has usefully labored to bridge the clash between east and west. An early illustration of Turkey's multilateralism and alliance-friendly role could be seen in the multiple crises accompanying Yugoslavia's break-up. Ankara pushed hard for UN intervention in the crisis and also sought to mediate between the Bosniak and Croat forces in Bosnia, which led to the Washington Agreement of 1994, establishing the Croat-Bosnian federation. Turkey also made its security forces available for UN, NATO, and indeed EU peacekeeping missions in former Yugoslavia. In April 1993, Turkey contributed to NATO's Operation Clear Skies, designed to impose a no-fly zone, and later contributed troops to UNPROFOR and to the follow-on NATO-led IFOR and SFOR. Ankara also contributed police officers to Bosnia, and its navy participated in Operation Sharp Guard. Later still, Turkey contributed around three hundred personnel to the EU Force (EUFOR-ALTHEA). Turkey subsequently contributed military and police assets to the NATO-led Kosovo Force (KFOR), to NATO and EU-led forces in Macedonia, and in Albania. At the time of writing there are in excess of one thousand uniformed Turkish personnel on peacekeeping missions in the Balkans, mostly in Kosovo. Turkey has been involved in training the armed forces of its Balkan friends, largely under the umbrella of the NATO PfP program, which has afforded Turkey a platform for the training of forces from the Caucasus and Central Asia. Indeed, NATO's PfP program and its peacekeeping activities have emerged as a significant vehicle for Turkish military and diplomatic outreach.

An additional illustration of Turkey's loyal ally behavior is offered by its contribution in Afghanistan in the wake of the October 2001 U.S.-led attack. Ankara was one of just eighteen NATO and PfP countries that contributed to the initial phase of the UN-mandated ISAF in December 2001. Turkey has subsequently twice taken command of ISAF, from June 2002 until February 2003 and from February to August 2005. Turkey has also headed the Kabul Regional Command.[23] In 2003, former Turkish Foreign Minister Hikmet Cetin was appointed as NATO's first Senior Civilian Representative in Afghanistan. In November 2006, Ankara established its own civilian-led PRT in

Wardak province, and in July 2010 Turkey opened a second PRT at Jowzan in the north of the country.

In addition to peacekeeping contributions in Somalia, Burundi, the Central African Republic, Chad, the Ivory Coast, Congo, Liberia, Sudan, Georgia, East Timor, and on the Iraq-Iran and Iraq- Kuwait borders, since 2006 Turkey has contributed assets to UNIFIL, and has made air bases and ports available to UNIFIL's contributors. In 2009, and in the wake of a number of pirate raids on Turkish merchant vessels, Ankara also decided to contribute to one of the four international maritime operations against Somalian pirates operating in the Gulf of Aden. Furthermore, Turkey put the base at Incirlik, vital in supporting the U.S. presence in Iraq, at the disposal of the United States.

Ankara's contribution to bridging differences is also reflected at the global level. Thus in February 2002, Ankara organized and hosted an EU/OIC conference as a counter to the divisive impact of the 9/11 incidents. The EU-OIC troika that emerged from the 2002 forum remains active. July 2005 saw the launch of the AoC initiative, cosponsored by Spain and Turkey. Its first forum was held in Madrid in January 2008 and the second in Istanbul in April 2009. The AoC launched a number of programs involving such diverse matters as youth employment, media initiatives, philanthropy, interreligious and cultural dialogue of various kinds, and the like. Although little has come of the initiative, it nevertheless offers additional testimony to Turkey's commitment to bridge divides and to desecuritize issues.[24]

Turkey's growing entanglement with the OIC might at first glance be seen as contradicting this general thrust and as supporting the thesis of an eastwards axis shift or "Islamification" of Turkish policy, but at closer inspection a different picture emerges. With the January 2005 elevation of a Turkish citizen, Ekmeleddin Ihsanoglu, as OIC Secretary General in the first ever election for that office, Turkey's influence within the OIC increased in line with its more active diplomacy toward the Islamic world. Ihsanoglu has spearheaded a reform effort aimed at extending the remit of the OIC, intensifying its focus on and embrace of global issues such as women's and human rights, and tightening the political relationships between its members. In 2005, a blueprint was laid down for a ten-year plan of action aimed at the promotion of moderation, modernization, reform, education, economic development, good governance, women's rights, and human rights in the Muslim world. Turkish efforts bore fruit with the OIC's adoption in 2008 of an updated charter that incorporated, among other innovations, an OIC human rights commission.

Of course, OIC member-states include some of the most repressive, corrupt, and economically undeveloped that the world has to offer. However, the Arab Awakening that erupted in the spring of 2011 might in due course enhance the possibility that Turkey might serve as a model for the democratization and development of the Muslim world. The idea that Turkish experience might provide a model for other Muslim majority states to emulate was initially seen as most applicable to the former Soviet Turkic states of central Asia and the Caspian region.[25] It received a boost in the wake of 9/11, when in association with the Bush administration's Broader Middle East and North Africa Initiative, some believed that the Turkish model might apply in the Arab world too. The AKP government's Foreign Minister Abdullah Gul in particular took up the theme, and he identified corruption, instability, and economic irrationality as among the problems facing Muslim societies and called for the Muslim world to strive toward better governance, transparency, accountability, gender equality, and improved human rights and freedoms.[26] Davutoglu has continued in a similar vein. The current popularity of the AKP government in the Middle East—a consequence of Erdogan's outspoken criticism of Israel, its growing economic role in the region, the impact of Turkish popular culture such as TV soap operas, and AKP-governed Turkey's apparent capacity to combine modernity with respect for Islamic values—has combined with recent developments in the Arab world to suggest that, for all its flaws, Turkey really could offer a source of inspiration for Arab populations and reformers.

Reflections: Will Turkey's New Approach to Security Last?

The Arab Awakening could prove to be just one of a number of factors that could force a rebalancing of Turkey's new security approach. Ankara has been torn between its close relationship with the region's regimes on the one hand and its sympathy toward popular demands for democratization on the other. It is possible that the region might in time move in Turkey's direction, but in the more immediate term, developments associated with this bout of turmoil in the Arab region have frequently left Ankara somewhat exposed. Turkey's economic fortunes in the region could also take a tumble, at least temporarily. Ankara is discovering that it cannot be a friend to all parties in a world containing so many rivalries. Furthermore, the Middle East's conservative regimes have sometimes been mistrustful of Turkey's friendships with the regions' more radical elements, and especially with non-Arab Iran. Nor have

Arab governments invariably appreciated non-Arab Turkey's intrusion into their domain. It remains far from clear that the Arab world is prepared to embrace Turkey, its former colonizer, as one of its own. It is noticeable that, in the spring of 2011, it was Egyptian rather than Turkish mediation that brought together the two Palestinian factions.

Examples of the risks entailed by Turkey's diplomacy of zero problems are mounting steadily. Turkey has lost its capacity to mediate between Israel and Syria, lost Israel's friendship, and has fallen out with Baghdad as a result of the Shia bid for power there, Iraq's support for Assad, Iranian influence, and Baghdad's dislike of Turkey's burgeoning friendship with the KRG. Indeed, the region's fissures have taken on an increasingly sectarian hue, and wittingly or otherwise Turkey has found itself cast as party of a Sunni bloc. Turkey's attempt to normalize relations with Armenia did not succeed, but it did arouse Azerbaijan's suspicions, which in turn could have implications for Ankara's ambition to develop as an energy hub. Ankara could also do little but stand on the sidelines while two other friends to its north, Russia and Georgia, went to war in 2008. Thereafter a degree of mistrust entered into Georgia's approach to Turkey. We should not decry Turkey's political and economic progress in the Balkans and in the former Soviet space, including Central Asia. However, the Balkan states aspire to join the EU, and the Caucasian and Central Asian states remain more in a Russian than a Turkish orbit. Some of the Central Asian states are tilting toward China, better placed both as market for the region's energy and as a source of investment and economic development. For all Davutoglu's astounding energy and the considerable progress he has made in raising Turkey's profile, tangible success is hard to locate. Turkey could find that, after all, it is not central but peripheral to the regions with which it overlaps, and that its neighbors will continue to look toward their fellow Arabs, or fellow Europeans, or toward Russia, rather than toward Turkey.

In the meantime, the AKP government has neglected, or at least marginalized as just one of many, its relationship with its Western allies. Combined with the slowing of domestic reform, this has not endeared the EU toward Turkish accession. Nor has there been much substantive progress on the Cyprus issue, for all the AKP government's initial courage, or with respect to Ankara and Athens. Turkey's energy relationship with, and dependence on, Russia has not enhanced its case to become the EU's energy partner of choice. Yet half of Turkey's trade is with, and the vast bulk of its inward investment is from, the EU. It is Turkey's largely EU-inspired democratization and economic development that makes Turkey attractive to the peoples of the Middle East and the wider Muslim world. Far more of Turkey's people live and work in Europe than elsewhere. Turkey's entanglement with Europe has both endured and deepened over time. This will not change for the foreseeable future.

Ankara's relationship with the United States raises even more imponderables than that with Europe. Turkey now has less need of U.S. economic or military assistance or protection than was formerly the case. It is evident too that U.S. policies in the region do not invariably offer a neat fit with Turkey's. However, the United States remains a crucial actor in each of the regions with which Turkey overlaps, and indeed globally. U.S. power and influence is a reality that Ankara must contend with. The Syrian uprising served to expose some of these contradictions in Turkish-U.S. relations. On the one hand, Ankara and Washington both sided with the rebels, and for a time the two governments moved closer to each other than had been the case for many years. On the other, Turkey became frustrated at its own inability to act in the absence of a degree of U.S. support and engagement that did not materialize.

Furthermore, future Turkish political leaders, if not the present incumbents of office, might in time revisit some of the recent tenets of Turkish foreign policy. Ankara's attempt to engage positively with Damascus and Tehran brought little return. They might better recognize the limits to Turkey's sway in Russia's sphere of influence. They might exhibit greater sensitivity to the mismatch between Turkish ambitions and energy on the one hand and its resources on the other. They might reconsider the degree to which Turkey's best interests are served by irritating its western allies. Turkey will not, cannot, and should not return to a Cold War dependence on the United States, or a more or less exclusive identification with Europe. Davutoglu is correct in asserting that Turkey is at one and the same time a European, Eurasian, Middle Eastern, economically developing, trading, democratizing, Islamic state and society. It should be unremarkable that its foreign policy might reflect that complexity. However, it is Turkey's interests that some of the inconsistencies, hypocrisies, over-ambition, over-confidence, and naiveties that Turkish policy has exhibited over the past few years should be ironed out. After all, de Gaulle's France was firmly located in an integrating Europe. But where is Turkey anchored, exactly?

Notes

1. Alan Makovsky, "The New Activism in Turkish Foreign Policy," *SAIS Review* 19(1), Winter-Spring 1999, pp. 92–113; Malik Mufti, "Daring and Caution in Turkish Foreign Policy," *Middle East Journal* 52(1), Winter 1998, pp. 32–50.

2. Ziya Onis, "The Turkish Economy at the Turn of the New Century," pp. 95–115, in Morton Abramowitz (ed.), *Turkey's Transformation and American Policy* (New York: Century Foundation Press, 2000); Mine Eder, "The Challenge of Globalization and Turkey's Changing Political Economy," pp. 189–215, in Barry Rubin and Kemal Kirisci (eds.), *Turkey in World Politics: An Emerging Multiregional Power* (Boulder, Colo., and London: Lynne Rienner, 2001).

3. James Ker-Lindsay, *Crisis and Conciliation: A Year of Rapprochement between Greece and Turkey* (London: I.B.Tauris, 2007).

4. Meliha Benli Altunisik and Ozlem Tur, "From Distant Neighbours to Partners? Changing Syrian-Turkish Relations," *Security Dialogue* 37(2), 2006, pp. 229–248; Bulent Aras and Hasan Koni "Turkish-Syrian Relations Revisited," *Arab Studies Quarterly* 24(4) Fall 2002, pp. 47–60.

5. William Hale, *Turkey, the US and Iraq* (London: SAQI, 2007), pp. 94–116; James E. Kapsis, "The Failure of U.S.-Turkish Pre-Iraq War Negotiations: An Overconfident United States, Political Mismanagement, and a Conflicted Military," *Middle East Review of International Affairs* 10(3), September 2006, 33–45; Bill Park, "Strategic Location, Political Dislocation: Turkey, the United States, and Northern Iraq," *Middle East Review of International Relations* 7(2), June 2003, pp.11–23; Michael Rubin, "A Comedy of Errors: American-Turkish Diplomacy and the Iraq war," *Turkish Policy Quarterly* 4(1), Spring 2005, pp. 69–79.

6. Bulent Aras, "Turkey's Rise in the Greater Middle East: Peacebuilding in the Periphery," *Journal of Balkan and Near Eastern Studies* 11(1), March 2009, pp. 29–41.

7. Ahmet Davutoglu, *Stratejik derinlik: Turkiye'nin uluslararasi konumu*, Istanbul: Kure Yayinlari, 2001. At the time of writing this book had not yet been translated into English.

8. For example, see his "Turkey's Foreign Policy Vision: An Assessment of 2007," *Insight Turkey* 10(1), 2008, pp. 77–96; "Turkish Foreign Policy and the EU," *Turkish Policy Quarterly*, 8(1), Fall 2009, pp. 11–17. For commentary on his thinking, see Bulent Aras, "The Davutoglu Era in Turkish Foreign Policy," *Insight Turkey* 11(3), Summer 2009, pp. 127–142; Ioannis N. Grigoriadis, "The Davutoglu Doctrine and Turkish Foreign Policy," Working Paper 8, Hellenic Foundation for European and Foreign Policy (ELIAMEP), April 2010. http://www.eliamep.gr/wp-content/uploads/2010/05/%CE%9A%CE%95%CE%99%CE%9C%CE%95%CE%9D%CE%9F-%CE%95%CE%A1%CE%93%CE%91%CE%A3%CE%99%CE%91%CE%A3-8_2010_IoGrigoriadis1.pdf, accessed 25 May 2011; and Alexander Murinson, "The Strategic Depth Doctrine of Turkish Foreign Policy," *Middle Eastern Studies*, 42(6), November 2006, pp. 945–964.

9. Omer Taspinar, "Turkey's Middle East Policies; between Neo-Ottomanism and Kemalism," Carnegie Papers 10, September 2008, http://www.carnegieendowment.org/files/cmec10_taspinar_final.pdf, accessed 25 May 2011.

10. Kemal Kirisci, "The Transformation of Turkish Foreign Policy: The Rise of the Trading State," *New Perspectives on Turkey* 40, 2009, pp. 29–57.

11. Tuncay Babali, "Turkey at the Energy Crossroads," *Middle East Quarterly*, XVI(2) Spring 2009 online version http://www.meforum.org/2108/turkey-at-the-energy-crossroad, accessed 5 October 2010; John Roberts, "Turkey as a Regional Energy Hub," *Insight Turkey* 12(3) Summer 2010, 39–48.

12. Semih Idiz, "The Turkish-Armenian Debacle," *Insight Turkey* 12(2), Spring 2010, pp. 11–19.

13. Henri J. Barkey, "Turkey's New Engagement in Iraq: Embracing Iraqi Kurdistan," *United States Institute of Peace Special Report* 237, May 2010.

14. Hakan Altinay, "Turkey's Soft Power: An Unpolished Gem or an Elusive Mirage?," *Insight Turkey*, 10(2), 2008, pp. 55–66; Meliha Benli Altunisik," The Possibilities and Limits of Turkey's Soft Power in the Middle East," *Insight Turkey* 10(2), 2008, pp. 41–54.

15. H. Akin Unver, "Turkey's 'Deep State' and the Ergenekon Conundrum," *The Middle East Institute Policy Brief*, no. 23, April 2009, http://www.mei.edu, accessed 25 May 2011; Umit Cizre and Joshua Walker, "Conceiving the New Turkey after Ergenekon," *International Spectator*, 45(1), 2010, pp. 89–98.

16. Ziya Onis, "Multiple Faces of the 'New' Turkish Foreign Policy: Underlying Dynamics and a Critique," *Insight Turkey*, 13(1), 2011, pp. 47–65.

17. Ilker Ayturk, "Between Crises and Cooperation: The Future of Turkish-Israeli Relations," *Insight Turkey*, 11(2), Spring 2009, pp. 57–74; Gokhan Bacik, "Turkish-Israeli Relations after Davos: A View from Turkey," *Insight Turkey* 11(2), Spring 2009, pp. 31–41; Ofra Bengio, "Altercating Interests and Orientations between Israel and Turkey: A View from Israel," *Insight Turkey* 11(2), Spring 2009, pp. 43–55; Taha Ozhan "Turkey, Israel and the U.S. in the Wake of the Gaza Flotilla Crisis," *Insight Turkey* 12(3), Summer 2010, pp. 7–18.

18. Bulent Aras and Rabia Karakaya Polat, "From Conflict to Cooperation: Desecuritization of Turkey's Relations with Syria and Iran," *Security Dialogue* 39(5), October 2008, pp. 495–515; Andrea Breitegger, "Turkish-Iranian Relations: A Reality Check," *Turkish Policy Quarterly* 8(3), Fall 2009, pp. 109–123; Daphne McCurdy, "Turkish-Iranian Relations: When Opposites Attract," *Turkish Policy Quarterly* 7(2), Summer 2008, pp. 87–106; Murat Mercan, "Turkish Foreign Policy and Iran," *Turkish Policy Quarterly* 8(4), Winter 2009–2010, pp. 13–19.

19. Nimrod Raphaeli, "The Growing Economic Relations between Iran and Turkey," The Middle East Media Research Institute (MEMRI) economic blog, 1 April 2008, http://www.memrieconomicblog.org/bin/content.cgi?article=92 accessed 28 May 2010.

20. Rahman G. Bonab, "Turkey's Emerging Role as a Mediator on Iran's Nuclear Activities," *Insight Turkey*, 11(3), Summer 2009, pp. 161–175.

21. Gregory L. Schulte, "Why Turkey Cannot Abstain on Iran's Nuclear Violations," *Turkish Policy Quarterly* 8(4), Winter 2009–2010, pp. 21–26.

22. Omer Taspinar, "The Rise of Turkish Gaullism: Getting Turkish American Relations Right," *Insight Turkey* 13(1), January-March 2011, pp. 11–17.

23. Hikmet Cetin, "Turkey's Role in Afghanistan," *Turkish Policy Quarterly* 6(2), Summer 2007, pp. 13–17; Lt. General Ethem Erdagi, "The ISAF Mission and Turkey's Role in Rebuilding the Afghan State," *Policy Watch* 1052, November 2005, http://www.washingtoninstitute.org/templateC05.php?CID=2403, accessed 6 September 2010; Nursin Atesoglu Guney, "The New Security Environment and Turkey's ISAF Experience," pp. 177–189, in Nursin Atesoglu Guney (ed.), *Contentious Issues of Security and the Future of Turkey* (Aldershot and Burlington, Vt.: Ashgate, 2007); Richard Weitz, "Turkey's Efforts to Support Afghanistan's Reconstruction," *Turkey Analyst*, 3(3), February 2010, http://www.silkroadstudies.org/new/inside/turkey/2010/100215B.html, accessed 6 September 2010.

24. Talha Kose, "The Alliance of Civilizations: Possibilities of Conflict Resolution at the Civilization Level," *Insight Turkey*, 11(3), Summer 2009, pp. 77–94.

25. Idris Bal, *Turkey's Relations with the West and the Turkic Republics: The Rise and Fall of the 'Turkish Model* (Aldershot, Vt.: Ashgate 2000); Andrew Mango, "The Turkish Model," *Middle Eastern Studies*, 29(4), 1993, pp. 726–757.

26. Abdullah Gul, "Turkey's Role in a Changing Middle East Environment," *Mediterranean Quarterly* 15(1), Winter 2004, pp. 1–7.

15 Conclusion

Andrew M. Dorman and Joyce P. Kaufman

In order to tie the pieces together, this chapter has been subdivided into three parts. The first examines each of the four groups of states in turn in order to consider what similarities and differences emerged amongst them, thus giving us an idea of the relative importance of history and structures within the international system to determine how states undertake national security formulation and implementation. The second part of the chapter then analyzes what factors can be identified as common across all thirteen case studies. Finally, the last section reflects on what wider considerations have emerged from this study that better help us understand the broader issues of national security policy.

The Old World

The case studies of France, Germany, and the United Kingdom (UK) provide an interesting insight into how the various European states are reflecting on national security and defense. The economic downturn is clearly in evidence with even states such as Germany, whose economy is based on a strong manufacturing base, finding that it is affected by the other European states. In particular, continuing doubts that surround the Euro have forced the German government to effectively underwrite the whole European economy. Doubts over the Greek, Portugese, Spanish, Italian, and now Cypriot economies have all forced German intervention and support for the European Central Bank.

This uncertainty looks set to continue with the recent Italian elections showing growing public support for the end of austerity measures. Thus the comedian cum-power broker Beppe Grillo has suggested that Italy move back to the Lira.[1] Interestingly, France's new president, Francois Hollande, appears to have backtracked on his opposition to austerity, raising taxes and imposing further public expenditure cuts.[2] The state of the European economy is clearly the principal security threat being identified by Europe's political elite even if public opinion is becoming less accepting of the austerity measures being imposed on Europe's economies. This is an entirely logical assumption; the debate concerns what action therefore needs to be taken and the potential impact that will have on Europe's position within the world.

The effect of this austerity is having a major impact on Europe's defense capabilities, with NATO's 2 percent of GDP to be spent on defense looking like a distant memory. Rumors surrounding the new French defense white paper (*livre blanc*) suggest that France may well follow the example of the UK and accept a step-change downwards in its power projection capabilities with hints of a budget closer to 1.3 percent of GDP plan. Further cuts to the UK's defense budget look certain on top of those announced in November 2012 by the Chancellor of the Exchequer.[3] This is being mirrored elsewhere and partly explains the German government's decision to end conscription. Thus we can see a Europe whose military forces are likely to become relatively weaker as investment in future capabilities continues to fall while existing forces are cut as part of the on-going austerity measures. This trend does not look like it will be reversed in the near future. The problem (if that really is the right expression) is that Europe has never been safer and has moved from being the continent that had the most wars and conflicts to the one with the least. Thus, public opinion in general does not see a need for spending significant amounts on its armed forces, especially when it is confronted with cuts to the welfare system.

Nevertheless, as Mali and Libya have shown, despite the continuing reductions and dependence on the United States for key capabilities, Europe's armed forces in some collective form still retain the ability to project military power beyond Europe. The real challenge is whether it retains the will to do so. At the political level in France and the UK, that seems to be the case and they remain the key lead nations. Moreover, the NATO-led operation over Libya also revealed a number of surprises in terms of who participated and who did not. Sweden, Belgium, Norway, and Denmark were prominent, with the former not even a member of NATO. Most commentators focused on the German absence. Perhaps more significant was the East/West split, with NATO's West European members involved while those to the East were noticeable in their absence, reflecting both a capability difference and also a difference in view over what NATO is principally about.

The 20th Century World

In this group, Australia, Canada, and Japan share many similarities with each other and also with the United States. All have seen their relative standing in the world grow after World War 2. All have aspirations of a permanent seat on the UN Security Council at some time, Canada because of its perceived good citizenship and Japan because of its relative economic might, although Australia's aspirations seem to have lessened. Nevertheless, all are now concerned about whether their relative power is beginning to wane and thus are looking to solidify their positions.

As a consequence, all three are concerned about their relationship with the United States and want to ensure that their defense and security relationship with the United States remains in place. Equally all have been identified by the United States as important partners, particularly because of geographical position. In the case of Australia and Canada, this includes the intelligence-sharing arrangements contained under the terms of the so-called Five Eyes agreement.

Whereas Canada has been a founding member of NATO, it is interesting to see that Japan and Australia's new partnerships with NATO reflect its growing global role and the wider security concerns of these states, and this NATO link has seen their respective forces serving under NATO command in Afghanistan. Nevertheless, for all three, the Pacific has become the main area of focus, although Canada is also increasingly concerned with the high north and the potential opening of the northern passageway. All three have retained advanced military capabilities and in recent years bucked the downward trend in relative defense spending amongst the advance Western states. However, austerity measures are also beginning to be seen in these three, and the planned levels of defense expenditure are being reduced. Most of this equipment has been acquired from the United States, although all three retain some indigenous capability. None have or seem intent on acquiring a nuclear capability, preferring to have this provided by the U.S. nuclear guarantee.

The (Re-)Emerging 21st Century World

There has been much academic discussion about the emergence or reemergence of the so-called BRIC nations (Brazil, Russia, India, and China). The three case studies examined (China, India, and Russia) revealed a number of similarities in their outlook and approach to national security. The first similarity is that all are very aware of each other and view each other as potential rivals or threats. This is perhaps not surprising in the India-China and the Russia-China cases because of their shared borders and the continuing border disputes, but it is also present in their relations with other nations, particularly over resources. Thus, their view of their relative rise is as much about their position with one another as other factors. They have all also sought to emphasize their position. Russian support for the Syrian government is, in part, about its regional interests and being seen to have an important role to play. The Chinese disputes with its neighbors over the South China Sea partially reflect its desire to show regional standing, while the increasing deployment of its navy further afield would seem to indicate that it wants to show that it will protect its interests wherever they are located. Interestingly, its involvement throughout Africa has, in part replaced that of the Soviet Union, albeit with a far greater resource focus. This also explains why the Indian government has sought to develop its maritime capabilities and project those further eastwards just as China pushes westwards.

Secondly, their individual relations with the United States are interesting with different strategies approached, in part reflecting the historical legacies of this relationship. Thus, Russia continues to want to be viewed as an equal of the United States as in the days of the Soviet Union and remains concerned with the growing in-balance between the two. For the United States, China is viewed as the main potential threat, and the Chinese government has sought to influence the United States through the acquisition of U.S. debt. However, the so-called pivot of U.S. forces toward the Pacific reflects U.S. concerns, and it looks as though of these rising nations China is the one most likely to end in some form of confrontation with the United States. In contrast, the U.S.–India relationship has transformed itself, and this is clearly a reflection of their mutual concern over China.

Thirdly, all three have secessionist movements within their current borders. This has raised a question mark about the homogeneity of them as nations, and internal unrest continues to be a source of tension and diversion.

Tibet, Kashmir, and Chechnya have not been resolved and look likely to remain unresolved for some time to come. Moreover, all three have potentially unstable countries on their borders, which could result in the displacement of people, and for Russia, threats to ethnic Russians living in these nations as a minority.

Fourthly, all three have significant demographic problems that will challenge them over the next few decades. China's one-child policy has given it a major imbalance between the sexes while India is scheduled to pass China as the world's largest country in terms of population size. Although it does not have the same one-child policy as China, India's cultural preference for male children has also resulted in an imbalance that has severe demographic implications for the future. Russia, in contrast, is suffering the problem of low birth rates, particularly amongst ethnic Russians, which will add to the pressures on the state in terms of areas such as health care.

Fifthly, all are investing in new military capabilities, and all struggle to match the technological capabilities of the United States while trying to develop their own indigenous programs. In the case of India, this has been partially eased by the increase recourse to external technologies such as U.S. and European aircraft. Nevertheless, all have difficult decisions ahead, and there remain question marks about their retention of armed forces on the current scale as they seek to invest more of their defense budgets on equipment rather than personnel.

The Potentially (Re-)Emerging States

The next potential generation of states that might follow on from the BRICs shares many similarities with them. Nigeria, Turkey, and South Korea all have either major internal challenges or, in the case of South Korea, a question mark over the reunification with North Korea. They are therefore very similar to the BRIC case studies within this volume. For all three, the internal challenges that confront them could very easily prevent their emergence as the next generation of Great Powers, and these interests will continue to dominate their thinking about national security.

Similarly, all have question marks raised about their identity—who do they want to be? In the case of Turkey, this is the debate between Europe and the Near East; for Nigeria it is its relative position within Africa; and for South Korea it is where the state fits within the Asia-Pacific region. Part of the problem for all three states is that the resolution of this question is not entirely in their hands. There are other powers within their region that will also influence their answers and which could again de-rail their relative advancement.

Nevertheless, all three have the potential to develop quite substantially. Although their economies are quite different, all three offer the potential to be regional economic hubs. Moreover, they have invested quite significantly in their armed forces, and all retain relative large militaries with a high level of sophistication in comparison with the area in which they are based. This means that their defense budgets are comparatively high, and whether they will be able to sustain this in the longer term will, in part, reflect local circumstances and also their desire for greater international engagement.

Interestingly, all three have a somewhat difficult relationship with the United States, which has ebbed and flowed over time. Partly, this is a reflection of the international environment and the degree to which the United States is focusing on the particular state, but it is also heavily influenced by domestic politics. For example, the change in power with Turkey away from the military has made it less pro-America and encouraged more localized security relationships.

Reflections across All the Case Studies

The first similarity that all the case studies reveal is that the elites within each of the states remain fixated about their states' relative standing internationally or even regionally, perhaps best illustrated by the British foreign secretary's emphasis on "no strategic shrinkage." How this emphasis is expressed differs amongst the states, partly because there is no universally agreed mechanism for articulating national security or even an agreement that such articulation is needed. Moreover, as part of this fixation there is also an emphasis on the states' trajectory, whether rising or falling, and relative position compared with traditional allies and rivals (e.g., India and China or France and the UK). In this respect policy-makers appear to assume that status and standing is a zero-sum game with little opportunity for mutual advancement. This would suggest that policy-makers are embracing some elements of classical realism whether consciously or subconsciously.[4]

Within this, there is a fixation on standing membership of the UN Security Council, which remains a consistent theme with states arguing over either the retention of their permanent standing (UK and France) or their suitability for permanent membership in the near term (e.g.. Japan, India, Brazil, Nigeria, and Germany). There were few exceptions (e.g., South Korea), but this was in part because they saw their interests covered by strategic partnership with the United States and recognized that their Chinese near neighbor was unlikely to allow such a move. This fixation is sometimes replicated within other organizations, most clearly within NATO and the jockeying over individual command appointments and the location of headquarters.

The articulation of national interest also varies between the states, but the concept of the national interest remains strong even though it, like national security, is an essentially contested concept. Its usage in various speeches by political leaders indicates that it continues to have both an international and a domestic context. In the case of the former, it represents a means of distinguishing between core and peripheral interests and thus serves as a useful means of diplomatic signaling. Its domestic context appears to serve two basic purposes. First, it allows state leaders to help justify their states' position or actions to the domestic audience. In other words, it forms a legitimizing function. Secondly, it provides a means of rallying domestic support for an individual state's actions and thus links directly to the Clausewitzian trinity of the state, the people, and the armed forces.[5] Here, like the concept of national security, it is an often-used phrase that assumes a common understanding that is not present. It was particularly noticeable amongst the "old world" states that the language of national interest has been used as an element of arguing for the preservation of the status quo, thus representing a reference to the past. Interestingly, the concept of national interest has also been broadened to incorporate values such as democracy in the case of the United States under President Bush or issues of ethnic cleansing in Prime Minister Tony Blair's "Doctrine of the International Community."

It is clear from the published national security documents that there has been a broadening of the security concept away from a purely military or defense focus. Indeed, looking at some of the documents, they contain a list of potential horrors confronting the respective state, both natural and man-made. Though the traditional understanding of national security is important in virtually all cases, we are also seeing increased attention to the broader understanding of security as it pertains to what has been referred to as human security. Few, however, attempt to provide any form of prioritization such as that outlined by the UK's most recent document, which stood out in contrast to its two predecessors. Instead, there appears to be general agreement that security now embraces all five elements outlined by Buzan in the early 1990s, with a number of the formal national security statements developing this further by embracing elements of the human security literature. At times, this appears a little contradictory, with the ideas of human security running against those of the nation state.

There is also general agreement that the relative balance of power with its Western dominance is increasingly being challenged, and references to a multipolar world have become more common. Quite a few public documents and far more speeches allude to the shift from a unipolar to a multipolar world in which the United States would no longer be dominant.[6]

There is evidence of a so-called Black Swan phenomenon with a clear difference between the speeches, statements, and published national security strategies that emerged before the 2008 financial crisis and those made subsequent to the crisis. For example, the Chinese 2008 defense paper made direct reference to the U.S. subprime mortgage collapse.[7] The 2008 crisis appears on initial examination to be exacerbating the relative changes in the balance of power with the European powers, in particular, and the United States to a lesser extent emphasizing that the financial crisis is the major security challenge to their states.

All the various reviews first emphasized the impact that globalization is having and the interdependency that this brings with it as the principal driver of change in the international system. None viewed this trend as reversible and all emphasized both the positive and negative elements of globalization including the increased interdependency of world trade, the overall loss of state control of the tools of violence, the rising threats posed by nonstate actors, and the greater vulnerability brought about by these changes in world trade. This internationalization is also reflected in a significant emphasis on the transatlantic alliance amongst the European and North American states examined most notably from both Germany and France.[8] The French white paper speaks of NATO and the EU being two aspects of a single approach and the importance of renewing its relations with NATO.[9] Clearly many states in the West wish to repair the damage that may have been done to their relations from previous decisions, and the transatlantic relationship is a core partnership in need of nurturing.[10]

The biggest problem for all these organizations is that their respective members disagree about the relative prioritization of challenges to the organization. For example, in NATO there are those states, principally the Eastern members, who view its principle function as the Article V mission against any form of revanchist Russia. By way of contrast, others, including the North American states, are focused on its global role.

It should also be stressed that there is no agreed or generally unified approach to reviewing defense or constructing a national security strategy or even one that can be associated with particular system of government (such as the Westminster model) or style of government (such as democratic). The majority of the states examined have either recently undertaken (2005 onwards) or are in the process of undertaking a defense review with the Australian 2009 white paper and Chinese biannual China's National Defense document for 2008 (published January 2009), being the first published within the context of the current world economic downturn, which the Australian government has sought to emphasize.[11] This message was picked up in the UK's 2010 Strategic Defence and Security Review.

There are a surprising number for whom this current defense white paper represents the first new public articulation of defense policy since the attacks on the United States in September 2001 (9/11) and the advent of the Global War on Terror (GWOT) (e.g., France, fourteen-year gap; Australia. nine years; Germany, twelve years; the Netherlands, nine years; New Zealand, nine years) although some have had periodic post-9/11 defense updates that attempted to maintain the currency of the previous pre–9/11 review. In this respect, many of the earlier defense reviews are contemporaries of the UK's 1998 Strategic Defence Review, which has clearly served as a model for some in terms of outlook and/or language, and the updates compare to the 2002 *SDR: A New Chapter.*[12]

Comparatively few states follow the U.S. system of a regularly mandated review, although a few, including the UK, have indicated that they intend to have a more regular review process than has subsequently occurred. For example, the Dutch are looking to produce a biannual strategic foresight report.[13] Interestingly, India and Japan have retained an approach in which an annual publication is produced (the UK abandoned this in 1999), whereas China has produced biannual defense policy documents. These are clearly on the fringes of what might count as a defense review but have been used as such in this paper.

Given that defense can be a highly politically charged issue within democratic states amongst those who have or are in the process of undertaking a major review, there have been a number of developments. First, there is a tendency in the current review round to create a review board involving quite a broad mix of individuals both from within and outside government to show openness. There is also a general move to obtain a political consensus and agree on a cross-party view. In France, for example, this involved a commission drawn from the armed forces, and defense and security establishments. More significantly, a much wider pool of expertise was invited to give their views including all the main political parties, academics, trade unions, members of the defense community, and the armed forces. Also significantly, the president sought to actively involve the French Parliament including members from the various parties on the commission and consulting the various parliamentary committees.[14] Likewise, the Australians sought to consult widely with a Community Consultation Panel conducting thirty public meetings, thirty-five private meetings, received over 450 written submissions, and got their costing independently verified.[15] Alternative approaches have tended to leave the analysis entirely to the defense establishment. In other words, there is an attempt to generally depoliticize defense and security policy. Such an approach inevitably is more time consuming, given the need to engage and be seen to engage outside government. Such an approach is not always practical given the current economic situation and the UK's Strategic Defence and Security Review was noticeable for its lack of engagement as the coalition government sought to plan its defense expenditure as part of a wider government spending review.

The level of engagement of a states' premier also varied with a fairly even split between defense ministers and premiers writing the foreword to the review. In general, it looks as though the apparently more contentious reviews had a foreword by the states' leader (e.g., President Sarkozy of France) as a way of showing support and reflecting that the defense policy ultimately is the responsibility of a states' leader within government.

There is also no agreed form of coordinating all the functions of state. A number of case studies have followed the U.S. example and formed a National Security Council, including most recently the UK, but there is little evidence to suggest that this has been any more successful than other forms of government that have a far looser arrangement.

The last decade has also witnessed the emergence of the first generation of public National Security Strategies (NSS).[16] A number of states have noticeably lagged behind in developing an NSS, with the UK's emerging finally in 2008 whereas the Canadians had an International Policy document in 2005 and the United States an NSS in 2002.[17] There were also some surprising omissions of public policy amongst some of the democratic states, in particular, Germany.[18] Consequently, the result has generally been a process that has seen a defense review published that focuses on defense's contribution to the delivery of defense and security, then an NSS is subsequently produced in some form that makes reference to defense but the level of integration is patchy. This is now being followed by the current round of defense reviews that generally sees a more joined-up approach linking the NSS to the defense policy, with the former document sitting above the defense policy, and from which the defense policy is developed. The reasoning for this convoluted approach has largely been bureaucratic, with the defense review process established within a particular department of state, and the more cross-governmental NSS process taking time to establish its position and the process of its development within the respective governmental system.

In outlining the current situation, all the reviews examined were unanimous in suggesting the world is a less safe place, although there is no uniform agreement about why the world is less safe or any measure to prove such a view. They all generally pointed to a series of trends that were undermining the ability of the state to protect or isolate itself from changes outside its borders. For example, the Chinese 2008 defense paper made direct reference to the U.S. subprime mortgage collapse.[19] While this probably represented a means of shifting blame for the economic downturn at home, it also showed an acknowledgment of the world's interdependence from a state that is less likely to acknowledge this.

Invariably, the second driver identified and linked to the first was the rise of terrorism, as in the words of the German white paper "a fundamental challenge and threat to freedom and security."[20] Although none went as far as the U.S. 2006 National Security Strategy and declared their country to be at war,[21] the 9/11 attacks on America were seen by many, in the words of the French white paper, to have "crossed an historical threshold and underwent a change of scale . . . ."[22] As a direct consequence, the majority of the reviews acknowledge either openly or tacitly that there has been a blurring in the traditional distinction between domestic and foreign security, and therefore, as a consequence, there is likely to be a far greater role for the armed forces in support of the civil administration at home.[24] Defense and support for the nation at home was a strong theme, with a number of reviews taking this further and directly identifying a role for the armed forces' role in coping with natural disasters at home and potentially abroad. This runs counter to the assumptions of the UK's 2003 and 2004 defense white papers, which sought to reduce the level of commitment at home in favor of short wars abroad.[24] In general, this shift appears to have been a mechanism for reinforcing a message to domestic public opinion of why they needed to support defense and their armed forces (i.e., their relevance was as much at home as abroad). The realities about their potential support for missions at home are therefore open to debate.

There was also an acknowledgment that there was potential for resource wars, with the Chinese paper using the most vociferous language, stating that "World peace and development are faced with multiple difficulties and challenges. Struggles for strategic resources, strategic locations and strategic dominance."[25] New areas of threat were also a constant theme. This was generally put in the form of an emphasis on asymmetric challenges. The definition of asymmetric challenges was generally quite loose and in some respects was more a catch-all term for other threats. Nevertheless, many of the papers identified cyberspace as a particular area of vulnerability from both potential state and nonstate opponents.[26] A few, most notably the United States, have also identified space as a potential environment for conflict. All the reviews examined emphasized the dangers posed by the proliferation of weapons of mass destruction and of the spread of ballistic missile technology. Consequently, all the reviews pledged to support the current international control regime in place. All the reviews also highlighted the potential dangers posed by nonstate actors, such as al Qaeda, with the French review being the most vociferous in this, arguing that the "privatization of armed violence is spreading" and implying that this might be reversed in some way.[27]

Only a few of the reviews make reference to climate change both in terms of the impact it will have and the impact on how the armed forces will operate in the future. In the former case, climate change is seen as a potential source of insecurity and in the Canadian example as a potential source of friction with its U.S. neighbor over access through a northwest passage. More generally, it is seen as a cause for population movements and the resultant destabilization that this might bring to various areas of the world. Linked to this is widespread concern about potential for conflict over resources and in the Chinese case strategic locations. The provision of future sources of power for the armed forces and the level to which they are able to contribute to global warming has only just begun to emerge as a potential issue.

Interestingly, two clear distinctions tend to separate this generation of reviews from their predecessors. First, there is the use of the terms defense and security rather than purely defense and the acknowledgment that defense contributes to the delivery of security. Some nations have managed to retain a more Soviet type of phraseology; for example, the Czech Republic's 2008 paper was entitled Military Strategy of the Czech Republic 2008.[28] The overall trend toward defense contributing to security rather than purely monopolizing the provision of defense is clear amongst the states examined.

Second, there is frequently an attempt to link defense policy to a national security strategy as part of what many now refer to as a comprehensive approach or integrated approach. Most are now trying to align these after initially publishing some form of NSS with a significant defense white paper already in existence and apparently unchanged.

As a consequence of these two distinctions, all the papers examined agree that the broader security threats now confronting the state mean that defense cannot provide for security alone and that it can only be a part of the overall response of government. There is, consequently, much discussion about cross-government approaches to defense and security both at home and abroad. The major difference with the UK is in defining what the role of defense and security is. They almost all speak of the protection of the state and its people. Thus, for the Dutch, "National security is under threat when vital interests of our society and/or state are threatened in such way that it leads to potential societal disruption."[29] Some states, such as Canada, have added the protection of the states' values, and others, such as France, its identity. In a sense, what both have tried to do is to aim to detect what makes them a nation. However, none have embraced the human security literature's views of a much broader security definition to one including the protection of the citizen from fear, which has been seen in the UK's 2009 National Security Strategy.[30] In other words, the state has a duty to protect itself, its citizens, and its way of life, but it does not have the responsibility of countering the fears that its citizens might have. Meaning, they tend to remain fairly traditional defense-type documents.[31]

All have individual linguistic idiosyncrasies that reflect the states' history, and this element does not change over time. Thus the French white paper speaks of "the capacity to guarantee the independence of France and the protection of all citizens."[32] Thus history and identity are important facets to individual papers. Nevertheless, the general thrust of change is fairly consistent among the democracies examined. Implicitly, and in some cases explicitly, the reviews contain a questioning of previous assumptions and policy options. For example, are such concepts as the ideas of deterrence and containment appropriate mechanisms for tackling the new asymmetric threats? One way has been to build on these earlier assumptions and incorporate them within a wider response. For example, the French have sought to build on their previous white papers. Their 1972 paper's central theme was deterrence, whereas the 1994 paper emphasized projection. The 2008 paper builds on these two and speaks of a balance between knowledge and anticipation, prevention, deterrence, protection, and intervention.[33] This then allows the white paper to articulate how the state will confront both conventional and unconventional threats.

In terms of the question of what is to be achieved, many adopt a concentric circle-type approach. Thus, the defense roles/missions speak of defense of the homeland, then defense of the region as they define which is often linked to alliances and then the wider world role with an emphasis on being a good citizen. Thus for Canada it is Canada, North America, the transatlantic partnership, and its international role. For Australia it is Australia, the immediate states close to it such as New Zealand and Papua New Guinea, then the Asia-Pacific region, and finally the wider world. This concentric circle-type approach is far closer to geographical focus of the UK's 1993 Statement on the Defence Estimates than on the thematic range of missions outlined in the 1998 Strategic Defence Review.[34] It also places an emphasis on home rather than international defense and security responsibilities.

Linked to this concentric circle approach, a number of the reviews speak about the importance of stabilization operations. The Canadians have argued in favor of the construction of Stabilization and Reconstruction Task Forces (START). In parallel to these moves is a clear trend toward putting limits on the potential level of support for such operations, with clear implications for Afghanistan. For example, the French speak of small contingents of 1–5,000[35] and the Danish 2004 paper pledged 2,000 personnel.[36] Thus internationalism is to be supported, but only to an acceptable level.

One consistent new term that is now being used is resilience—the need for the state and its people to be able to sustain them in the event of a natural or man-made catastrophic incident, such as Hurricane Katrina or the 9/11 attacks. Most speak of resilience in terms of a national response, but a number of states that share borders with other nations, such as the Dutch, have emphasized that "[c]ountries are dependent on each other if they wish to increase their resilience."[37] In other words, isolationism is not possible, as the movement of people and animals cannot be controlled.

In terms of outlook, the various defense and national security strategy documents generally all look forward for between fifteen and twenty years. The Australians, for example, project to 2030, the Canadians look ahead twenty years to 2026, France to 2025, the Dutch fifteen to twenty years, and the Americans are mandated to look ahead twenty years. Few try to project as far as the UK's Strategic Trends document, and this raises interesting questions for measuring relative success for areas such as soft power.[38] The challenge will be in sustaining such reviews implementation, and it is not clear what processes will be adopted. One way may be to provide periodic updates along the lines that the Australian and New Zealanders did over the last decade.

The accompanying pledges on defense spending do not, however, generally look ahead that far. All the reviews published so far, except the Australian defense review, were before the financial crisis that currently confronts the world. Unlike the UK, in general there is a commitment to maintaining defense funding at its current levels or increasing defense spending in real terms in the years ahead. Thus France has pledged to increase the defense budget year-on-year in real terms by 1 percent from 2012,[39] and the Canadians pledged $13 billion over twenty years to help restore their armed forces. Here the Canadians, along with the United States, have been somewhat of an exception and began to look at the cost of reconstituting their force post-Afghanistan (and in the U.S. case, post-Iraq). However, in the United States, the Congressional-imposed sequester will have a direct impact on defense spending. The others remain, at least in public, quiet on this front.

In general, the international and domestic constraints tend to be implicit rather than explicitly stated. Legitimacy for actions remains important, and a number of reviews speak of intervention-type operations only under the authority of a UN resolution and frequently only as part of a coalition. In other words, there is a domestic requirement for legitimacy and international top cover. Whether this is a temporary reaction to the conflicts in Iraq and Afghanistan or whether it is a more permanent move is unclear. All call for the international community to come together to deal with the challenges they have identified, even though some argue that collective response is weakening.

A few speak of managing risk, acknowledging that the state can no longer (if it ever could) provide for the entire protection of the nation, and this implies that economic constraints will play a part. This links to the concentric circle idea, with the circles closest to the heartland being the most vigorously protected and the softer support to the international community being more opaque.

Not surprisingly, there is division between those engaged in Afghanistan and those who are not. Of those engaged in operations in southern Afghanistan (Canada, the Netherlands, and Australia), there is an implied reluctance to extend their existing commitment and a refocusing on homeland defense, yet there is also a significant commitment from these countries to defense spending with pledges quite a way forward. Perhaps the most surprising element is the lack of articulation in those engaged in Afghanistan. None of the reviews discuss the implications of success or failure in Afghanistan for either the state concerned, NATO, or the wider Western alliance. This appears, perhaps not surprisingly, to be a taboo area for discussion, which would be understandable.

If this section had been written prior to NATO's commencement of operations over Libya in support of the UN Security Resolution 1973 or the French deployment to Mali, the answer would appear to have been an increasing reluctance to use force. However, the Libyan operation and now Mali appear to run counter to many of the arguments set out in various states' national security strategies and defense policy documents. A European-led NATO mission in which the United States took a backseat from offensive operations has led to a reemergence, in the language of "Liberal Interventionism," at least in the UK, and by implication the suggestion that these so-called cosmopolitan values may still have some standing.

That said, it is clear that states are focusing on the economic situation, at least in the short term, with the result that within the "old world" and soon in the 20th century world defense expenditure is being significantly reduced and the interrelationship of economics and the price of defense are becoming increasingly apparent.

Wider Reflections

It is very easy for academics to argue that the international system is at a so-called tipping point, but it is clear that the relative balance of power and individual states' trajectories within this remain a prime focus for national leaders. We may well be seeing a significant shift in the balance of power away from the West exacerbated by the economic downturn, the cost of the various conflicts of the first decade of the 21st century, and the West's desire to maintain its relative standard of living. How this will unfold in future is even less clear, with an emphasis placed on maintaining the status quo.

It is also interesting to note that the language of national security remains essentially realist in its articulation, and that there appears to have been few inroads from the other international relation's theoretical traditions. With the exception of occasional references to human security, the dialogue is a mix of classical and neo-realism, reflecting the dominance of this school both within U.S. academia and policy-makers more generally. How this is altered is debatable and partly a reflection of the other schools' disinterest in national security. That said, liberalism's emphasis on cooperation and integration is reflected by the strong emphasis on the almost universal assumption that globalization will inevitably continue and deepen. In this respect, therefore, national security remains dominated by both the realist and liberalist paradigms.

Finally, national security remains largely measured in terms of so-called hard power and in particular, on military capabilities. Although there is a debate about the future direction of conflict and therefore the types of capability that a state may need, there remains uniform agreement that all states should acquire and maintain a military capability, and that the more significant a state is, the greater this capability should be.

Notes

1. James Legge, "Beppe Grillo Suggests Iyaly May Leave the Euro," *The Independent Online*, 3 March 2013, http://www.independent.co.uk/news/world/europe/beppe-grillo-suggests-italy-might-leave-the-euro-8517790.html accessed 4 March 2013.

2. John Lichfield, "France gets a €30bn Austerity from Francois Hollande U-turn," *The Independent Online*, 11 September 2012, http://www.independent.co.uk/news/world/europe/france-gets-a-30bn-austerity-shock-from-franois-hollande-uturn-8122406.html, accessed 4 March 2013.

3. James Kirkup, "Defence Spending Cut Again," *Telegraph Online*, 5 December 2012, http://www.telegraph.co.uk/news/politics/9725591/Defence-spending-cut-again.html, accessed 4 March 2013.

4. See, for example, E. H. Carr, *The Twenty Years' Crisis 1919–1939: An Introduction to the Study of International Relations* (London: Macmillan, 1939); Hans J. Morgenthau, *Politics among Nations: The Struggle for Power and Peace*, 5th ed., revised (New York: Alfred Knopf, 1978); John Mearsheimer, "Structural Realism," in Timothy Dunne, Milja Kurki, and Steve Smith, eds., *International Relations Theories: Discipline*

and Diversity (Oxford: Oxford University Press, 2006); Kenneth N. Waltz, *Man, the State and War: A Theoretical Analysis* (New York: Columbia University Press, 1964).

5. See Colin S. Gray, *The Strategy Bridger: Thoeory for Practice* (Oxford: Oxford University Press, 2010); Carl von Clausewitz, *On War* (London: Penguin Books, 1964).

6. "China's National Defense in 2008," Beijing, 2009, p. 3, http://merln.ndu.edu/whitepapers/China_English2008.pdf.

7. "China's National Defense in 2008," Beijing, 2009, p. 4, http://merln.ndu.edu/whitepapers/China_English2008.pdf.

8. Federal Ministry of Defence, "White Paper 2006—on German Security Policy and the Future of the Bundeswehr," 2006, pp. 29–45,

9. "The French White Paper on Defence and National Security," 2008, p. 93, http://merln.ndu.edu/whitepapers/France_English2008.pdf.

10. See Andrew M. Dorman and Joyce P. Kaukman (eds.), *The Future of Transatlantic Relations: Perceptions, Policy and Practice* (Palo Alto, Calif.: Stanford University Press, 2011).

11. "Defending Australia in the Asia Pacific Century: Force 2030," Australian Department of Defence, 2009, http://www.defence.gov.au/whitepaper/docs/defence_white_paper_2009.pdf; "China's National Defense in 2008," Beijing, 2009, http://merlin.ndu.edu/whitepapers/China_English2008.pdf both accessed 1 June 2010.

12. Ministry of Defence, "The Strategic Defence Review," *Cm.3,999* (London: TSO, 1998), http://www.mod.uk/NR/rdonlyres/65F3D7AC-4340–4119–93A2–20825848E50E/0/sdr1998_complete.pdf accessed 18 November 2008; Ministry of Defence, "SDR: A New Chapter," *Cm.5,566*, (London: TSO, 2002), http://www.mod.uk/NR/rdonlyres/79542E9C-1104–4AFA-9A4D-8520F35C5C93/0/sdr_a_new_chapter_cm5566_vol1.pdf accessed 22 March 2008.

13. "National Security Strategy and Work programme 2007–8," Ministry of the Interior and Kingdom Relations, The Hague, 2007, p. 10.

14. "The French White Paper on Defence and National Security," 2008, p. 10, http://merln.ndu.edu/whitepapers/France_English2008.pdf.

15. "Defending Australia in the Asia Pacific Century: Force 2030," Australian Department of Defence, 2009, http://www.defence.gov.au/whitepaper/docs/defence_white_paper_2009.pdf. p.18.

16. Cabinet Office, "The National Security Strategy of the United Kingdom: Security in an interdependent world," *Cm.7,291* (London: TSO, 2008), http://interactive.cabinetoffice.gov.uk/documents/security/national_security_strategy.pdf accessed 21 March 2008.

17. "The National Security Strategy of the United States" (Washington, D.C.: White House, 2002), http://merln.ndu.edu/whitepapers/USnss2002.pdf accessed 4 June 2010.

18. See Nicholas Floyd, "How Defence Can Contribute to Australia's National Security Strategy," (Sydney: Lowy Institute for International Policy, 2009).

19. "China's National Defense in 2008," Beijing, 2009, p. 4, http://merln.ndu.edu/whitepapers/China_English2008.pdf.

20. Federal Ministry of Defence, "White Paper 2006—on German Security policy and the Future of the Bundeswehr," 2006, p. 5.

21. President George W. Bush, "The National Security Strategy of the United States of America," March 2006, http://merln.ndu.edu/whitepapers/USnss2006.pdf.

22. "The French White Paper on Defence and National Security," 2008, p. 27, http://merln.ndu.edu/whitepapers/France_English2008.pdf.

23. See, for example, the foreword by President Sarkozy, "The French White Paper on Defence and National Security," 2008, p. 9, http://merln.ndu.edu/whitepapers/France_English2008.pdf

24. "SDR: A New Chapter," *Cm.5,566* (London: TSO, 2002), http://www.mod.uk/NR/rdonlyres/DD89DBE6-CEAA-4995-9E01-52EF6D19FC73/0/sdr_a_new_chapter_cm5566_vol2.pdf accessed 22 March 2008; "Delivering Security in a Changing World: Defence White Paper," *Cm.6.041-I* (London: TSO, 2003), http://www.mod.uk/NR/rdonlyres/051AF365-0A97-4550-99C0-4D87D7C95DED/0/cm6041I_whitepaper2003.pdf; "Delivering Security in a Changing World: Future Capabilities," *Cm.6,269* (London: TSO, 2004). Both accessed 4 June 2010.

25. "China's National Defense in 2008," Beijing, 2009, p. 3, http://merln.ndu.edu/whitepapers/China_English2008.pdf.

26. See Paul Cornish, "Cyber Security and Politically, Socially and Religiously Motivated Cyber Attacks" (London: Chatham House, 2009), http://www.chathamhouse.org.uk/files/13346_0209_eu_cybersecurity.pdf; Paul Cornish et. al., "Cyberspace and the National Security of the United Kingdom: Threats and Responses" (London: Chatham House, 2009), http://www.chathamhouse.org.uk/files/13679_r0309cyberspace.pdf both accessed 4 June 2010.

27. "The French White Paper on Defence and National Security," 2008, p. 29, http://merln.ndu.edu/whitepapers/France_English2008.pdf.

28. "Military Strategy of the Czech Republic 2008," Prague, 2008, http://merln.ndu.edu/whitepapers/Czech_Republic_English-2008.pdf

29. "National Security Strategy and Work programme 2007–8," Ministry of the Interior and Kingdom Relations, The Hague, 2007, p. 11.

30. Cabinet Office, "The National Security Strategy of the United Kingdom: Update 2009: Security for the Next Generation," *Cm.7,590* (London: TSO, 2009), http://www.cabinetoffice.gov.uk/media/216734/nss2009v2.pdf accessed 8 October 2009.

31. Federal Ministry of Defence, "White Paper 2006—on German Security Policy and the Future of the Bundeswehr," 2006, p. 55,

32. "The French White Paper on Defence and National Security," 2008, p. 9, http://merln.ndu.edu/whitepapers/France_English2008.pdf.

33. "The French White Paper on Defence and National Security," 2008, p. 60, http://merln.ndu.edu/whitepapers/France_English2008.pdf.

34. "Defending Our Future: Statement on the Defence Estimates, 1993," *Cm.2,270* (London, HMSO, 1993).

35. "The French White Paper on Defence and National Security," 2008, p. 122, http://merln.ndu.edu/whitepapers/France_English2008.pdf.

36. "The Danish Defence Agreement 2005—2009," 2004, p. 4, http://merln.ndu.edu/whitepapers.html

37. "National Security Strategy and Work programme 2007–8," Ministry of the Interior and Kingdom Relations, The Hague, 2007, p. 10.

38. "Global Strategic Trends out to 2040," 4th ed. (London: DCDC, 2010), http://www.mod.uk/NR/rdonlyres/38651ACB-D9A9–4494–98AA-1C86433BB673/0/gst4_update9_Feb10.pdf accessed 4 June 2010.

39. The French White Paper on Defence and National Security," 2008, p. 278, http://merln.ndu.edu/whitepapers/France_English2008.pdf.

Bibliography

Abramowitz, Morton (ed.), *Turkey's Transformation and American Policy* (New York: Century Foundation Press, 2000).

Altinay, Hakan, "Turkey's Soft Power: An unpolished Gem or an Elusive Mirage?," *Insight Turkey*, vol. 10, no. 2, 2008, pp. 55–66.

Altunisik, Meliha Benli, "The Possibilities and Limits of Turkey's Soft Power in the Middle East," *Insight Turkey*, vol. 10, no. 2, 2008, pp. 41–54.

Altunisik, Meliha Benli, and Tur, Ozlem, "From Distant Neighbours to Partners? Changing Syrian-Turkish Relations," *Security Dialogue*, vol. 37, no. 2, 2006, pp. 229–248.

Appadorai, A., *Select Documents on India's Foreign Policy and Relations 1947–1972, Vol. 1* (Oxford: Oxford University Press, 1982).

Aras, Bulent, "The Davutoglu Era in Turkish Foreign Policy," *Insight Turkey*, vol. 11, no. 3, Summer 2009, pp. 127–142.

Aras, Bulent, "Turkey's Rise in the Greater Middle East: Peacebuilding in the Periphery," *Journal of Balkan and Near Eastern Studies*, vol. 11, no. 1, March 2009, pp. 29–41.

Aras, Bulent, and Koni, Hasan, "Turkish-Syrian Relations Revisited," *Arab Studies Quarterly*, vol. 24, no. 4, Fall 2002, pp. 47–60.

Aras, Bulent, and Polat, Rabia Karakaya, "From Conflict to Cooperation: Desecuritization of Turkey's Relations with Syria and Iran," *Security Dialogue*, vol. 39, no. 5, October 2008, pp. 495–515.

Asmussen, Mikkel Vedby, *The Risk Society: Terror, Technology and Strategy in the Twenty-First Century* (Cambridge: Cambridge University Press, 2006).

Armstrong, David, *The Rise of the International Organisation: A Short History* (Basingstoke, UK: Macmillan, 1982).

Art, Robert J., and Waltz, Kenneth N., *The Use of Force* (Lanham, Md.: Rowman & Littlefield Publishers Inc., 1999).

Axelrod, Robert, *The Evolution of Cooperation* (New York: Basic Books, 1984).

Axelrod, Robert, and Keohane, Robert O., "Achieving Cooperation under Anarchy: Strategies and Institutions," *World Politics*, 38(1), October 1985.

Ayturk, Ilker, "Between Crises and Cooperation: The Future of Turkish-Israeli Relations," *Insight Turkey*, vol. 11, no. 2, Spring 2009, pp. 57–74.

Babali, Tuncay, "Turkey at the Energy Crossroads," *Middle East Quarterly*, vol. XVI, no. 2, Spring 2009.

Bacik, Gokhan, "Turkish-Israeli Relations after Davos: A View from Turkey," *Insight Turkey*, vol. 11, no. 2, Spring 2009, pp. 31–41.

Badescu, Christina, "Canada's Continuing Engagement with United Nations Peace Operations," *Canadian Foreign Policy*, vol. 16, no. 2, 2010.

Bal, Idris, *Turkey's Relations with the West and the Turkic Republics: The Rise and Fall of the "Turkish Model"* (Aldershot, UK: Ashgate 2000).

Barkey, Henri J., "Turkey's New Engagement in Iraq: Embracing Iraqi Kurdistan," *United States Institute of Peace Special Report 237*, May 2010.

Bechtol, Bruce E. (Jr.), "Preparing for Future Threats and Regional Challenges: The ROK-U.S. Military Alliance in 2008–09," *Shifting Strategic and Political Relations with the Koreas*, Joint U.S.-Korea Academic Studies, vol. 19 (Washington, D.C.: Korea Economic Institute, 2009), pp. 75–99.

Bell, Coral, *Living with Giants: Finding Australia's Place in a More Complex World* (Canberra, Australia: ASPI, April 2005).

Bengio, Ofra, "Altercating Interests and Orientations between Israel and Turkey: A View from Israel," *Insight Turkey*, vol. 11, no. 2, Spring 2009, pp. 43–55.

Bernstein, George L., *The Myth of Decline: The Rise of Britain Since 1945* (London: Pimlico, 2004).

Blackwill, Robert D., and Dibb, Paul (eds.), *America's Asian Alliances* (Cambridge, Mass.: MIT Press, 2000).

Blair, Tony, *A Journey* (London: Hutchinson, 2010).

Bock, P. G., and Berkowitz, Morton, "The Emerging Field of National Security," *World Politics*, vol. 19, 1966, pp. 122–136.

Bonab, Rahman G., "Turkey's Emerging Role as a Mediator on Iran's Nuclear Activities," *Insight Turkey*, vol. 11, no. 3, Summer 2009, pp. 161–175.

Booth, Ken, "Security and Emancipation," *Review of International Studies*, vol. 17, no. 4, 1991, pp. 313–326.

Booth, Ken, and Smith, Steve (eds.), *International Relations Theory Today* (Cambridge, UK: Polity Press, 1995).

Booth, Ken, and Todd, Russell (eds.), *Strategic Cultures in the Asia-Pacific Region* (Basingstoke, UK: Macmillan, 1999).

Breitegger, Andrea, "Turkish-Iranian Relations: A Reality Check," *Turkish Policy Quarterly*, vol. 8, no. 3, Fall 2009, pp. 109–123.

Brown, Chris, *Understanding International Relations* (Basingstoke, UK: Palgrave, 2001).
Brown, Michael E., Lynn-Jones, Sean M., and Miller, Steven E., *Debating the Democratic Peace* (Cambridge, Mass.: MIT Press, 1996).
Brown, Michael E., Lynn-Jones, Sean M., and Miller, Steven E., *Theories of War and Peace* (Cambridge Mass.: MIT Press, 1998).
Brown, Michael E., et al. (eds.), *Rational Choice and Security Studies: Stephen Walt and His Critics* (London: MIT Press, 2000).
Bull, Hedley, *The Anarchical Society* (London: Macmillan, 1995).
Bull, Hedley, and Watson, Adam (eds.), *The Expansion of International Society* (Oxford, UK: Clarendon Press, 1984).
Bulley, Dan, "'Foreign' Terror? London Bombings, Resistance and the Failing State," *British Journal of Politics and International Relations*, vol. 10, no. 3, August 2008. pp. 379–394.
Buzan, Barry, *People, States and Fear: An Agenda for International Security Studies in the Post-Cold War Era* (London: Longman, 1991).
Carr, E. H., *The Twenty Years' Crisis 1919–1939: An Introduction to the Study of International Relations* (London: Macmillan, 1939).
Cetin, Hikmet, "Turkey's Role in Afghanistan," *Turkish Policy Quarterly*, vol. 6, no. 2, Summer 2007, pp. 13–17.
Chapnick, Adam, "The Middle Power," *Canadian Foreign Policy*, vol. 7, no. 2, Winter 1999.
Cho, Lee-Jay, Ahn, Chung-si, and Kim, Choong-nam (eds.), *A Changing Korea in Regional and Global Contexts* (Honolulu and Seoul: East-West Center and Seoul National University Press, 2004).
Clunan, Anne L., *The Social Construction of Russia's Resurgence: Aspirations, Identity and Security Interests* (Baltimore: Johns Hopkins University Press, 2009).
Cornish, Paul N. and Dorman, Andrew M., 'Blair's Wars and Brown's Budgets: From Strategic Defence Review to Strategic Decay in Less than a Decade," *International Affairs*, vol. 85, no. 2, March 2009, pp. 247–261.
Cornish, Paul N. and Dorman, Andrew M., "National Defence in the Age of Austerity," *International Affairs*, vol. 85, no. 4, July 2009, pp. 733–753.
Cornish, Paul N. and Dorman, Andrew M., "Dr Fox and the Philosopher's Stone: The Alchemy of National Defence in the Age of Austerity," *International Affairs*, vol. 87, no. 2, March 2011, pp. 335–353.
Croft, Stuart, *Culture, Crisis and America's War on Terror* (Cambridge: Cambridge University Press, 2006).
Croft, Stuart, and Terriff, Terry (eds.), *Critical Reflections on Security and Change* (London: Frank Cass, 2000).
Dawisha, Karen, and Parrott, Bruce, *Russia and the New States of Eurasia: The Politics of Upheaval* (Cambridge, Mass.: Cambridge University Press, 1994).
Deutsch, Karl, *Political Community in the North Atlantic Area* (Princeton, New Jersey: Princeton University Press, 1957).

Donaldson, Robert H., and Nogee, Joseph L., *The Foreign Policy of Russia: Changing Systems, Enduring Interests*, 4th ed. (Armonk, N.Y.: M.E. Sharpe, Inc., 2009).

Drew, Elizabeth, *On the Edge: The Clinton Presidency* (New York: Simon & Schuster 1994).

Dorman, Andrew M., *Blair's Successful War: British Military Intervention in Sierra Leone* (Farnham, UK: Ashgate, 2009).

Dorman, Andrew M., "Reorganising the Infantry: Drivers of Change and What This Tells Us about the State of the Defence Debate Today," *British Journal of Politics and International Relations*, vol. 8, no. 4, 2006, pp. 489–502.

Dorman, Andrew M., "Making 2 + 2 = 5: The 2010 Strategic Defence and Security Review," *Defense and Security Analysis*, vol. 27, no. 1, March 2011, pp. 77–87.

Dorman, Andrew M., and Kaufman, Joyce P. (eds.), *The Future of Transatlantic Relations: Perceptions, Policy and Practice* (Stanford, Calif.: Stanford University Press, Security Series, 2010).

Dunne, Tim, *Inventing International Society* (London: Macmillan, 1998).

Dupont, Alan, "Australia and the Concept of National Security," Strategic and Defence Studies Centre, *ANU Working Paper no. 206*, May1990.

Dutta, Sujit, "Managing and Engaging Rising China: India's Evolving Posture," *Washington Quarterly*, vol. 34, no. 2, 2011, pp. 127–144.

Edwards, Charlie, *National Security for the Twenty-First Century* (London: DEMOS, 2007).

Elias, Juanita, and Carol Johnson, "On Re-engaging Asia," *Australian Journal of Political Science*, vol. 45, no. 1, March 2010, pp. 1–12.

Elman, Colin, and Elman, Miriam (eds.), *Progress in International Relations Theory: Appraising the Field* (Cambridge, Mass.: MIT Press, 2003).

Fang, Yang, "China's New Marine Interests: Implications for Southeast Asia," *RSIS Commentaries*, no. 9/7/2011 (July 4, 2011).

Fierke, K. M., *Critical Approaches to International Security* (Cambridge, UK: Polity Press, 2007).

Fitz-Gerald, Ann M., "A UK National Security Strategy: Institutional and Cultural Challenges," *Defence Studies*, 8(1), March 2008, pp. 4–25.

Flemes, Daniel, "Regional Power South Africa: Co-Operative Hegemony Constrained Historical Legacy," *Journal of Contemporary African Studies*, vol. 27, no. 2, April 2009.

Floyd, Nicholas, "How Defence Can Contribute to Australia's National Security Strategy," *Lowy Institute for International Policy*, Sydney, 2009.

Fukuyama, Francis, *The End of History and the Last Man* (Harmondsworth, UK: Penguin, 1992).

Gaffney, John, "Highly Emotional States: French-US Relations and the Iraq war," *European Security*, vol. 13 no. 3 (part one), 2004.

Gaidar, Yegor, *Collapse of an Empire: Lessons for Modern Russia*, trans. Antonina W. Bouis (Washington, D.C.: Brookings Institution Press, 2007).

Gallie, W. B., "Essentially Contested Concepts." *Proceedings of the Aristotelian Society*, vol. 56, 1955–1956.

Ganguly, Sumit (ed.), *Indian Foreign Policy: Retrospect and Prospect* (New Delhi: Oxford University Press, 2010), pp. 206–225.

Gelb, Leslie H., "GDP Now Matters More than Force: A U.S. Foreign Policy for the Age of Economic Power," *Foreign Affairs*, vol. 89, no. 6, Nov.-Dec. 2010).

Ghazvinian, John, *Untapped: The Scramble for Africa Oil* (Orlando, Florida and London: Harcourt Inc, 2007).

Goldstein, Avery, *Rising to the Challenge: China's Grand Strategy and International Security*, (Stanford, Calif.: Stanford University Press, 2005).

Gordon, Philip H., *Winning the Right War: The Path to Security for America and the World* (New York: Times Books, 2007).

Guney, Nursin Atesoglu (ed.), *Contentious Issues of Security and the Future of Turkey* (Aldershot, UK and Burlington, Vt.: Ashgate, 2007).

Haas, Richard N., *War of Necessity, War of Choice: A Memoir of Two Iraq Wars* (New York: Simon & Schuster, 2009).

Haftendorn, H., et al. (eds.), *The Strategic Triangle: France, Germany, and the United States in the Shaping of the New Europe* (Johns Hopkins U. Press and Woodrow Wilson Center Press, 2006).

Harnisch, Sebastian, and Maull, Hanns W., *Germany as a Civilian Power? The Foreign Policy of the Berlin Republic* (Manchester, UK: Manchester University Press, 2001).

Harrison, Brian, *Seeking a Role: The United Kingdom 1951–70* (Clarendon Press: Oxford, 2009).

Herspring, Dale, "Russia's Crumbling Military," *Current History*, October 1998.

Hill, J.N.C, *Nigeria since Independence: Forever Fragile?* (Basingstoke, Hampshire and New York: Palgrave Macmillan, 2012).

Hirst, Christian, "The Paradigm Shift: 11 September and Australia's Strategic Reformation' *Australian Journal of International Affairs*, vol. 61, no. 2, June 2007, pp. 175–192.

Hogan, Michael J., *A Cross of Iron: Harry S Truman and the Origins of the National Security State, 1945–1954* (New York: Cambridge University Press, 1998).

Hoskins, Andrew, and O'Loughlin, Ben, *Television and Terror: Conflicting Times and the Crisis of News Discourse* (Basingstoke: Palgrave Macmillan, 2007).

Howorth, Jolyon, "Sarkozy and the 'American Mirage' or Why Gaullist Continuity Will Overshadow Transcendence," *European Political Science*, vol. 9, no. 2, 2010.

Hughes, Christopher W., *Japan's Re-emergence as a "Normal" Military Power* (Oxford, Oxford University Press, 2004).

Hui, Zhang, "Space Weaponization and Space Security: A Chinese Perspective," *China Security*, Issue 2, 2006.

Huntington, Samuel P., *The Third Wave: Democratization in the Late Twentieth Century* (Norman: Oklahoma University Press, 1991).

Huth, Paul K., and Allee, Todd L. *The Democratic Peace and Territorial Conflict in the Twentieth Century* (Cambridge: Cambridge University Press, 2002).

Idiz, Semih, "The Turkish-Armenian Debacle," *Insight Turkey*, vol. 12, no. 2, Spring 2010, pp. 11–19.

Indyk, Martin S., Lieberthal, Kenneth G., and O'Hanlon, Michael E., *Bending History: Barack Obama's Foreign Policy* (Washington, D.C.: The Brookings Institution, 2012).

Jennings, Peter, "Developing Actionable Intelligence for Homeland Security," *Australian Strategic Policy Institute*, 13 July 2005.

Jinming, Li, and Dexia, Li, "The Dotted Line on the Chinese Map of the South China Sea," *Ocean Development & International Law*, vol. 34 (London: Taylor & Francis Inc., 2003).

Jisi, Wang, "China's Search for a Grand Strategy: A Rising Great Power Finds Its Way," *Foreign Affairs*, vol. 90, no. 2, March/April 2011, pp. 68–79.

Jones, Ben, "Franco-British Cooperation: A New Engine for European Defence?," *European Union Institute for Security Studies Occasional Paper 88*, February 2011.

Kagan, Robert, *The World America Made* (New York: Alfred A. Knopf, 2013).

Kapsis, James E., "The Failure of US-Turkish Pre-Iraq War Negotiations: An Overconfident United States, Political Mismanagement, and a Conflicted Military," *Middle East Review of International Affairs*, vol. 10, no. 3, September 2006, pp. 33–45.

Kapur, S. Paul, and Ganguly, Samut, "The Transformation of US-India Relations: An Explanation for the Rapprochement and Prospects for the Future," *Asian Survey*, vol. 47, no. 4, pp. 642–656.

Katzenstein, Peter (ed.), in *The Culture of National Security: Norms and Identity in World Politics*, (New York: Columbia University Press, 1996).

Keating, Tom, "Multilateralism: Past Imperfect, Future Conditional," *Canadian Foreign Policy*, vol. 16, no. 2, 2010.

Kennedy, Paul, *The Rise and Fall of Great Powers* (New York: Vintage Books, 1987).

Keohane, Robert O., and Nye, Joseph S., *Power and Interdependence* (New York: Longman, 2000).

Ker-Lindsay, James, *Crisis and Conciliation: A Year of Rapprochement between Greece and Turkey* (London: I.B.Tauris, 2007).

Klare, Michael T., *Blood and Oil: The Dangers and Consequences of America's Growing Petroleum Dependency* (London and New York: Penguin Books, 2004).

Knorr, Klaus, *Historical Dimensions of National Security Problems* (Lawrence: University of Kansas, 1976).

Kolodziej, Edward A., *Security and International Relations* (Cambridge: Cambridge University Press, 2005).

Koo, Yongnok, and Suh, Dae-sook (eds.), *Korea and the United States: A Century of Cooperation* (Honolulu: University of Hawaii Press, 1984).

Kose, Talha, "The Alliance of Civilizations: Possibilities of Conflict Resolution at the Civilization Level," *Insight Turkey*, vol. 11, no. 3, Summer 2009, pp. 77–94.

Krause, Joachim, "Multilateralism: Behind European Views," *The Washington Quarterly*, vol. 27, no. 2, 2004.

Lansford, Tom, "Whither Lafayette? French Military Policy and the American Campaign in Afghanistan," *European Security*, vol. 11, no. 3, 2002, pp. 136–137.

Layne, Christopher, "The Unipolar Illusion: Why New Great Powers Will Rise?" *International Security*, vol. 17, no. 4, Spring 1993, pp. 9–10.

Ledwidge, Frank, *Losing Small Wars: British Military Failure in Iraq and Afghanistan* (New Haven, Conn.: Yale University Press, 2011).

Lee, Sook-jong, "Allying with the United States: Changing South Korean Attitudes," *The Korean Journal of Defense Analysis*, vol. 17, no. 1, Spring 2005, pp. 81–104.

Liping, Xia, "The New Security Concept in China's New Thinking of International Security," *International Review*, vol. 34, Spring 2004, pp. 29–42.

Little, Richard J., *The Balance of Power in International Relations: Metaphors, Myths and Models*. (Cambridge: Cambridge University Press, 2007).

McCourt, David, "Rethinking Britain's *Role in the World* for a New Decade: The Limits of Discursive Therapy and the Promise of Field Theory," *British Journal of Politics and International Relations*, vol. 13, no. 2, May 2011, pp. 145–164.

McCurdy, Daphne, "Turkish-Iranian Relations: When Opposites Attract," *Turkish Policy Quarterly*, vol. 7, no. 2, Summer 2008, pp. 87–106.

McMaster, H. R., "On War: Lessons to be Learned," *Survival*, vol. 50, no. 1, February-March 2008.

McMillan, James F., *Twentieth Century France: Politics and Society 1898–1991* (London: Edward Arnold, 1992).

Mahbubani, Kishore, *The New Asian Hemisphere: The Irresistible Shift of Global Power to the East* (New York: Public Affairs, 2008).

Makovsky, Alan, "The New Activism in Turkish Foreign Policy," *SAIS Review*, vol. 19, no. 1, Winter-Spring 1999, pp. 92–113.

Mandelbaum, Michael, *The Fates of Nations: The Search for National Security in the Nineteenth and Twentieth Centuries* (Cambridge: Cambridge University Press, 1988).

Mango, Andrew, "The Turkish Model," *Middle Eastern Studies*, vol. 29, no. 4, 1993, pp. 726–757.

Marr, Andrew, *A History of Modern Britain* (London: Pan Books, 2007).

Mearsheimer, John J, "The Gathering Storm: China's Challenge to U.S. Power in Asia," *The Chinese Journal of International Politics*, vol. 3, 2010.

Menon, Anand, with Dimitrakopoulos, D., and Passas, A., "France and the EU under Sarkozy: Between European Ambitions and National Objectives?," *Modern & Contemporary France*, vol. 18, no. 4, 2009.

Menon, Anand, "Empowering Paradise? ESDP at Ten," *International Affairs*, vol. 85, no. 2, March 2009.

Menon, K.P.S, *Many Worlds* (New Delhi: Oxford University Press, 1965).

Mercan, Murat, "Turkish Foreign Policy and Iran," *Turkish Policy Quarterly*, vol. 8, no. 4, Winter 2009–2010, pp. 13–19.

Miller, Benjamin, *States, Nations and the Great Powers: The Sources of Regional War and Peace* (Cambridge: Cambridge University Press, 2007).

Mingjiang, Li, "Reconciliing Assertiveness and Cooperation? China's Changing Approach to the South China Sea Dispute," *Security Challenges*, vol. 6, no. 2, Winter 2010, pp. 49–68.

Moon, Chung-in, and Li, Chun-fu, "Reactive Nationalism and South Korea's Foreign Policy on China and Japan: A Comparative Analysis," *Pacific Focus*, vol. 25, no. 3, 2010, pp. 331–355.

Morgenthau, Hans J, *Politics among Nations: The Struggle for Power and Peace*, 5th ed., revised (New York: Alfred Knopf, 1978).

Morse, Edward L., *Foreign Policy and Interdependence in Gaullist France* (Princeton, N.J.: Princeton University Press, 1973).

Morse, Eric S., "Analysis of the Obama Administration's National Security Strategy 2010," *The National Strategy Forum Review*, 3 June 2010.

Mufti, Malik, "Daring and Caution in Turkish Foreign Policy," *Middle East Journal*, vol. 52, no. 1, Winter 1998, pp. 32–50.

Murinson, Alexander, "The Strategic Depth Doctrine of Turkish Foreign Policy," *Middle Eastern Studies*, vol. 42, no. 6, November 2006, pp. 945–964.

Nam, Joo-Hong, *America's Commitment to South Korea: The First Decade of the Nixon Doctrine* (Cambridge: Cambridge University Press, 1986).

North, Richard, *Ministry of Defeat: The British War in Iraq 2003–2009* (London: Continuum UK, 2009).

Nye, Joseph S., *The Paradox of American Power: Why the World's Only Superpower Can't Go It Alone* (Oxford: Oxford University Press, 2003).

Nye, Joseph S., *Soft Power: The Means to Success in World Politics* (New York: Public Affairs, 2005).

Nye, Joseph S., "The Future of American Power: Dominance and Decline in Perspective," *Foreign Affairs*, Nov./Dec. 2010.

Onis, Ziya, "Multiple Faces of the 'New' Turkish Foreign Policy: Underlying Dynamics and a Critique," *Insight Turkey*, vol. 13, no. 1, 2011, pp. 47–65.

Ozhan, Taha, "Turkey, Israel and the US in the Wake of the Gaza Flotilla Crisis," *Insight Turkey*, vol. 12, no. 3, Summer 2010, pp. 7–18.

Pant, Harsh V., *Contemporary Debates in Indian Foreign and Security Policy: India Negotiates Its Rise in the International System* (New York: Palgrave Macmillan, 2008).

Pant, Harsh V. (ed.), *Indian Foreign Policy in a Unipolar World* (London: Routledge, 2009).

Pant, Harsh V., *The China Syndrome: Grappling with an Uneasy Relationship* (New Delhi: HarperCollins, 2010).

Pant, Harsh V., *China's Rising Global Profile: The Great Power Tradition* (Portland, Ore.: Sussex Academic Press, 2011).

Pant, Harsh V., *The US-India Nuclear Pact: Policy, Process and Great Power Politics* (Oxford: Oxford University Press, 2011).

Park, Bill, "Strategic Location, Political Dislocation: Turkey, the United States, and Northern Iraq," *Middle East Review of International Relations*, vol. 7, no. 2, June 2003, pp. 11–23.

Park, Changhee, "Why China Attacks: China's Geostrategic Vulnerability and Its Military Intervention," *Korean Journal of Defense Analysis*, vol. 20, no. 3, 2008, pp. 263–282;

Paul, T. V., Wirtz, James J., and Fortmann, Michel (eds.), *Balance of Power: Theory and Practice in the 21st Century* (Stanford, Calif.: Stanford University Press, 2004).

Pérouse de Montclos, Marc-Antoine, "Maritime Piracy in Nigeria: Old Wine in New Bottles?," *Studies in Conflict and Terrorism*, 2012, vol. 35, nos. 7–8.

Powell, Colin L, "U.S. Forces: Challenges Ahead," *Foreign Affairs*, Winter 1992/93.

Przystup, James, "The U.S.-Japan Alliance and the U.S.-ROK Alliance," *Shifting Strategic and Political Relations with the Koreas, Joint U.S. Korea Academic Studies*, vol. 19 (Washington, D.C.: Korea Economic Institute, 2009), pp. 66–67.

Purdum, Todd S., *A Time of Our Choosing: America's War in Iraq* (New York: Times Books, 2003).

Ralph, Jason G., *Beyond the Security Dilemma: Ending America's Cold War* (Aldershot, UK: Ashgate, 2001).

Rice, Condoleezza, "Promoting the National Interest," *Foreign Affairs*, vol. 79, no. 1, Jan–Feb. 2000).

Rittberger, Volker (ed.), *German Foreign Policy since Unification, Issues in German Politics* (Manchester, UK: Manchester University Press, 2001).

Roberts, John, "Turkey as a Regional Energy Hub," *Insight Turkey*, vol. 12, no. 3, Summer 2010, pp. 39–48.

Ross, Robert S., and Feng, Zhu (eds.), *China's Ascent: Power, Security, and the Future of International Politics* (Ithaca. N.Y.: Cornell University Press, 2008).

Roubini, Nouriel, "China Guesses Wrong on Growth," *Australian Financial Review*, 18 April 2011.

Rubin, Barry, and Kirisci, Kemal (eds.), *Turkey in World Politics: An Emerging Multiregional Power*, Boulder, Colo., and London: Lynne Rienner, 2001).

Rubin, Michael, "A Comedy of Errors: American-Turkish Diplomacy and the Iraq War," *Turkish Policy Quarterly*, vol. 4, no. 1, Spring, 2005, pp. 69–79.

Sanger, David E., *Confront and Conceal: Obama's Secret Wars and Surprising Use of American Power* (New York: Crown Publishers, 2012).

Sanger, David E., *The Inheritance: The World Obama Confronts and the Challenges to American Power* (New York: Three Rivers Press, 2009).

Schulte, Gregory L., "Why Turkey Cannot Abstain on Iran's Nuclear Violations," *Turkish Policy Quarterly*, vol. 8, no. 4, Winter 2009–2010, pp. 21–26.

Searle, John, *The Construction of Social Reality* (London: Allen Lane, 1995).

Serfaty, Simon, "Moving into a Post-Western World," *Washington Quarterly*, vol. 34, no. 2, 2011, pp. 7–23.

Shambaugh, David, "Coping with a Conflicted China," *The Washington Quarterly*, vol. 34, no. 1, Winter 2011, pp. 7–27.

Sheehan, Michael, *The Balance of Power: History and Theory* (London: Routledge, 1996).

Sheehan, Michael, *International Security: An Analytical Survey* (Boulder, Colo.: Lynne Rienner, 2005).

Shin, Gi-wook, *One Alliance, Two Lenses: US-Korea Relations in a New Era* (Stanford. Calif.: Stanford University Press, 2010).

Simon, Sheldon W., "Theater Security Cooperation in the U.S. Pacific Command: An Assessment and Projection," *NBR Analysis*, vol.14, no. 2, August 2003.

Smith, Jordan Michael, "Reinventing Canada: Stephen Harper's Conservative Revolution," *World Affairs*, March/April 2012, pp. 23–24.

Smith, Martin A., *Russia and NATO since 1991: From Cold War through Cold Peace to Partnership?* (London and New York: Routledge, 2006).

Snyder, Jack, "One World, Rival Theories," *Foreign Policy*, Nov./Dec. 2004.

Strachan, Hew, "The Strategic Gap in British Defence Policy," *Survival*, vol. 51, no. 4, August–September 2009, pp. 49–70.

Subrahmanyam, K., *Indian Security Perspectives* (New Delhi: ABC Publishing House, 1982).

Swanstrom, Niklas, "Security and Conflict Management in East Asia," *The Korean Journal of Defense Analysis*, vol. 20, no. 3, 2008, pp. 187–198.

Taeho, Kim, "Sino-ROK Relations at a Crossroads: Looming Tensions amid Growing Interdependence," *The Korean Journal of Defense Analysis*, vol. 17, no. 1, 2005, pp. 129–149.

Taspinar, Omer, "The Rise of Turkish Gaullism: Getting Turkish American Relations Right," *Insight Turkey*, vol. 13, no. 1, January-March 2011, pp. 11–17.

Tellis, Ashley J. et al., *Asia Responds to Its Rising Powers* (Washington, D.C.: The National Bureau of Asian Research, 2011), pp. 101–128.

Tellis, Ashley J., *India's Emerging Nuclear Posture: Between Recessed deterrent and Ready Arsenal* (New York: Oxford University Press, 2001).

Treacher, Adrian, *French Interventionism: Europe's Last Global Player?* (Aldershot, UK: Ashgate, 2003).

Ungerer, Carl, "The Intelligence Reform Agenda: What Next?," *Australian Strategic Policy Institute*, 27 February 2008.

Ungerer, Carl, "Connecting the Docs: Towards an Integrated National Security Statement," *Australian Strategic Policy Institute,* 10 December 2009.

Ungerer, Carl, "Measuring up: Evaluating Cohesion in the National Security Community," *Australian Strategic Policy Institute,* June 2010.

Ungerer, Carl, "Australia's National Security Institutions: Reform and Renewal," *Australian Strategic Policy Institute,* September 2010.

Ungerer, Carl, and Anthony Bergin, "The Devil in the Detail: Australia's First National Security Statement," *Australian Strategic Policy Institute,* 10 December 2008.

Unver, H. Akin, "Turkey's 'Deep State' and the Ergenekon Conundrum," *The Middle East Institute Policy Brief,* no. 23, April 2009.

Wall, Irwin M., "The French-American War over Iraq," *Brown Journal of World Affairs*, vol. X, issue 2, Winter/Spring 2004.

Walt, Stephen, *The Origins of Alliances* (Ithaca, N.Y.: Cornell University Press, 1987).

Walt, Stephen, "International Relations: One World, Many Theories," *Foreign Policy*, Spring 1998.

Waltz, Kenneth N., *Man, the State and War: A Theoretical Analysis* (New York: Columbia University Press, 1964).

Waltz, Kenneth N., *Theories of International Politics* (Reading, Mass.: Addison-Wesley, 1979).

Waltz Kenneth N., *Realism and International Politics* (New York: Routledge, 2008).

Watson, Adam, *The Evolution of International Society: A Comparative Historical Analysis* (London: Routledge, 2002).

Wendt, Alexander, "Anarchy Is What States Make of It," *International Organization*, vol. 46, no. 2, Spring 1992, pp. 391–425.

Wendt, Alexander, "Constructing International Politics," *International Security*, vol. 20, no. 1, Summer 1995, pp. 71–81.

White, Hugh, *The China Choice: Why America Should Share Power* (Collingwood, Australia: Black Inc., 2012).

White, Hugh, "Powershift: Australia's Future between Washington and Beijing," *Quarterly Essay*, 39, 2010.

Wight, Martin, *Power Politics* (London: Pelican Books, 1979).

Wight, Martin, *Systems of States* (Leicester, UK: Leicester University Press, 1977).

Williams, Philip M., and Harrison, Martin, *De Gaulle's Republic* (London: Longmans, 1961).

Wolfers, Arnold, "'National Security' as an Ambiguous Symbol," *Political Science Quarterly*, vol. 67, no. 4, December 1952.

Yost, David S, "France's New Nuclear Doctrine," *International Affairs*, vol. 82, no. 4, July 2006, pp. 701–721.

Zakaria, Fareed, *The Post-American World* (New York: W.W. Norton & Co., 2008).

Zenko, Micah, *Between Threats and War: U.S. Discrete Military Operations in the Post Cold War World*, Council on Foreign Relations, 2010.

Index